The Search for Solutions

HORACE FREELAND JUDSON

The Search for Solutions

ABRIDGED EDITION
(text complete, pictures and captions omitted)

Introduction by Lewis Thomas, M.D.

1987
The Johns Hopkins University Press
BALTIMORE

Printed in the United States of America

The abridgment of this Johns Hopkins Paperbacks edition of *The Search for Solutions* is limited to the deletion of photographs and captions from the original hardcover edition. The text of the book has not been altered.

The illustrated edition was published by Holt, Rinehart and Winston.
This edition is reprinted by arrangement with Henry Holt and Company, Inc.

The Johns Hopkins University Press
701 West 40th Street
Baltimore, Maryland 21211

Library of Congress Cataloging-in-Publication Data

Judson, Horace Freeland.
The search for solutions.

Bibliography: p.
Includes index.
1. Science—Social aspects. 2. Technology—Social aspects.
3. Scientists—Interviews. I. Title.
Q175.5.J82 1987 500 87-2856
ISBN 0-8018-3526-7 (pbk.)

Photo credits: p. 21, World Wide Photos; p. 50, Mike Jackson; p. 66, Susan Shaw; pp. 90, 100, Playback Associates; p. 153, Bachrach; p. 182, Caltech Photo; p. 193, Wayne Miller/Magnum Photos; p. 244, courtesy Florida State University

. . . still for Tom

Contents

Preface to the Johns Hopkins Edition: Knowledge and Power

The Search for Solutions, to define the book in a sentence, is an introduction to epistemology for the general reader—which is to say, it is about how we know—lightly written, full of anecdotes and examples from sciences of the past and of the present day, serious in intent. Scientific knowledge, collectively, is the most reliable knowledge we have got. The aim of the book is to ransack the enormous disparity of the sciences for instances that display some underlying unities of approach that bring us to such reliable knowledge.

The approaches of scientists and the knowledge they generate are more than, and somewhat different from, the facts and explanations reported from week to week in scientific periodicals and the press and codified periodically in textbooks. This would seem a truism, except that the reasons for it are not widely appreciated. To begin with, the content of sciences—what gets reported and eventually appears in textbooks—is inseparable from processes of discovery and testing. The sciences are always in motion. More, the content of the sciences is interconnected, a vast set of networks, densely woven and knotted. Facts and explanations—data, regularities, laws, theories low level and high—are not significant in isolation but appear at the multiple intersections of the networks. In principle, every one of these intersections and knots is always subject to the tug of individual test, and to rejection or revision.

This is obviously the case with new scientific knowledge, the recent observations and explanations whose accuracy and significance are still in the initial phase of evaluation. The place of new things in the multiple intersections can take a while to determine. Just last March, for example, a British mathematician, Colin Rourke, and his Portuguese colleague, Eduardo Rego, announced that they had achieved a proof for an idea known as Poincaré's conjecture. The conjecture itself is simple enough, though we don't need the details here, and is fundamental to much other mathematics: mathematicians will be deeply troubled if it be not true. Yet proof of it has eluded them for the eighty years since Henri Poincaré suggested it. The claim of a proof was intensely exciting, at least to the trade. But Rourke and Rego's proof is said to be exceptionally hard to evaluate. In the months since their announcement, doubts have grown. At a conference in November at the University of California in Berkeley, other mathematicians pointed out a gap in the proof; Rourke acknowledged that the gap will be hard to bridge and that he might have to withdraw the claim. As I write, the outcome is not known.

We expect such things at the boundaries as they are extended. But even matters in the heartland of science, matters long thought settled, sometimes come under new scrutiny. This year, for example, George S. K. Wong, of the National Research Council of Canada, pointed out that an exceedingly elementary, well-established value—the speed of sound in the atmosphere under defined standard temperature and pressure—had been miscalculated. His correction was small, a twentieth of one per cent, yet potentially of importance to aerodynamics and acoustics. Also just this year, a more fundamental and more firmly established matter came into question, namely, Galileo's assertion, in the seventeenth century, that all bodies, no matter what they are made of, are accelerated by the earth's gravitation at the same rate. That was last tested directly in experiments carried out between 1899 and 1908 by a Hungarian physicist, Roland von Eötvös, who compared, with an accuracy of one part in a billion, the pull of the earth's gravity on various materials. Eötvös's experiments had been thought definitively to verify Galileo's assertion. But now Ephraim Fischbach and colleagues say that they have reexamined Eötvös's data and conclude, though tentatively, that another force, weak and previously unknown, may operate on certain materials but not others, depending on the composition of the atoms' nuclei, to repel even as gravitation attracts. Other scientists have begun planning further tests. As experiment and opinion have developed thus far, the revisionist suggestion is not winning adherents and Galileo's assertion seems secure. But here, too, the outcome is not yet known.

If not just the new but even the most secure doctrines of science can become unsettled, where is the reliability of scientific knowledge? The answer is, of course, *right there.* The reliability of the sciences is akin to the reliability of an immense and complex piece of machinery whose builders are running it even as they build it: every part of the sciences being subject to inspection and test, the entire apparatus—those interconnected networks—is resilient and strong. The unique confidence we have in the sciences is based not on renewed faith but on constant, vigilant, most searching skepticism.

Knowledge is power, wrote Francis Bacon, four hundred years ago, even at the origin of modern science. But did he? We all know the phrase and think we know what it promises—and how abundantly the promise has been fulfilled since Bacon offered it. In the first chapter of this book, as it was originally published, I quoted Bacon's phrase, and too confidently went on to explain it: "Bacon was not a scientist. He wrote as a bureaucrat in retirement. His slogan was actually the first clear statement of the promise by which, ever since, bureaucrats justify to each other and to king or taxpayer the spending of money on science. Knowledge is power; today we would say, less grandly, that science is essential to technology."

Alas, there I committed a crude misreading (and misdating) of what Bacon said. To be sure, Bacon had an inspired vision—his acquaintance Christopher Marlowe might have understood it as a Faustian vision—of the future of science. Yet that line of Bacon's proposed a different covenant. What Bacon wrote was, rather, "*Nam et ipsa scientia potestas est.*" Which is to say, "In and of itself, knowledge is power." The difference? He was not making, not here anyway, Mephistopheles' vulgar claim that knowledge facilitates power, that dominion over the world can be wrung from knowledge as a practical consequence. Nor was he dividing pure science from applied science: his Renaissance *scientia* was not closely synonymous with today's "science," and anyway he would have thought the division superficial. Bacon was asserting something that these days seems subtle: that knowledge is power in its own right—because knowledge is knowledge.

The distinction is pointed up by the context, routinely ignored, in which Bacon wrote that sentence. It does not appear in one of his great treatises on the method and prospects of science. It was not, after all, a Faustian statement. You will find it in an early work, from the outset of his career, his *Meditationes Sacrae,* or *Religious Meditations,* specifically his essay "On Heresies." Bacon here prefigured, if anything, not the hydrogen bomb but the *Areopagitica*—Milton's defense of freedom of the press, in matters of belief and doctrine, with the assertion that not the censorship but this freedom is truth's sword and shield. Knowledge is power because it entails right thinking. That is, to know something—really to know it, not merely know *of* it—is to think in a particular way and, necessarily, not to think in certain other ways. Still more, such knowledge is not passive, it does not know things "idlie and as a looker on" (so Bacon translated his own Latin a few sentences later), but is unitary with right action.

In short, the deepest attraction of the sciences lies not in the resulting technology but in the sciences themselves, above all in their ways of knowing the world and showing it to us.

The Search for Solutions was first published in 1980, in a format that was elaborately designed and profligately illustrated in full color. It sold well but was not issued in soft cover because the production costs were judged to be too high. The idea of publishing it in soft cover without the copious color illustrations and their captions was first suggested several years ago by Marian Wood, the book's editor at Henry Holt and Company; this is what the Johns Hopkins University Press is now doing. This edition corrects the misreading of Francis Bacon just discussed, and a few factual and typographical errors. For it I have written, besides this preface, a revised bibliography. Though the pictures in the first edition were gorgeous, this version without them is, I think, actually a better book. Free of distraction, the aim and the argument emerge more distinctly.

29 December 1986

Introduction by Lewis Thomas, M.D.

The transformation of human society by science is probably only at its beginning, and nobody can guess at how it will all turn out. At the moment, the most obvious and visible effects on our lives are those resulting from the technology that derives from science, for better or worse, and much of today's public argument over whether science is good or bad is really an argument about the value of the technology, not about science. Walking around on the moon was a feat of world-class engineering, made possible by two centuries of classical physics, most of it accomplished by physicists with no faintest notion of walking on the moon. Penicillin was a form of technology made feasible by sixty years of fundamental research in bacteriology (you had to have streptococci and staphylococci in hand, and you had to know their names as well as their habits, before you could begin thinking of such things as antibiotics). Nuclear bombs, nuclear power plants, and radioisotopes for the study of human disease symbolize the range of society's choices for the application of mid-twentieth-century science, as were in their day the light bulb, the automobile, and the dial telephone.

But technology is only one aspect of science, perhaps in the very long run the least important. Quite apart from the instruments it has made available for our survival, comfort, entertainment, or annihilation, science affects the way we think together.

The effects on human thought tend to come at us gradually, often subtly and sometimes subconsciously. Because of new information, we change the ways in which we view the world. We are not so much the center of things anymore, because of science. We feel lost, or at least not yet found. We are not as informed about our role in the universe as we thought we were a few centuries ago, and despite our vast population we live more in loneliness and, at many times, dismay. Science is said to be the one human endeavor above all others that we should rely on for making predictions (after all, prediction is a central business of science), yet, for all the fact that we live in the Age of Science, we feel less able to foretell the future than ever before in our history.

This is, to date, the most wrenching of all the transformations that science has imposed on human consciousness—in the Western world, anyway. We have learned that we do not really understand nature at all, and the more information we receive, the more strange and mystifying is the picture before us. There was a time, just a few centuries back, when it was not so. We thought then that we could comprehend almost all important matters: the earth was the centerpiece of the universe, we humans were the centerpiece of the earth, God was in his heaven just beyond what we have now identified as a narrow layer of ozone, and all was essentially right with the world; we were in full charge, for better or worse.

Now, we know different, or think we do. There is no center holding anywhere, as far as we can see, and we can see great distances. What we thought to be the great laws of physics turn out to be local ordinances, subject to revision any day. Time is an imaginary space. We live in a very small spot, and for all we know there may be millions of other small spots like ours in the millions of other galaxies; in theory, the universe can sprout life any old time it feels like it, anywhere, even though the other parts of our own tiny solar system have turned out to be appallingly, depressingly dead. The near views we have had of Mars, and what we have seen of the surfaces of Jupiter and Venus, are a new cause of sadness in our culture; humans have never before seen, close up, so vast a lifelessness. It is, when you give it a thought, shocking.

It is sometimes made to seem that the sciences have already come most of their allotted distance and we have learned most of what we will ever know. Lord Kelvin is reported to have concluded as much for physics, near the turn of this century, with an announcement that physics was a finished, perfected discipline with only a few odds and ends needing tidying up; soon thereafter came X rays, quantum theory, and relativity—and physics was back at the beginning again.

I believe this is the true nature of science, and I can imagine no terminal point of human inquiry into nature, ever. Biology has been undergoing explosive activity during the past quarter-century; we call it the Biological Revolution because we have learned so much, so fast. But it is clear that biology is only starting up. What we perceive, so far, is order and beauty, and a kind of perfection in every reductionist detail of life, but also an increasing sense of strangeness. There is a long, long way to go before we comprehend life, and maybe we never shall. The three deepest mysteries are the evolution of the whole, coherent system of life in which we are working parts; the sorting out of cells and tissues in embryos; and our own brains. I am not sure we can ever solve the last one, for there must always be a kind of uncertainty associated with the effort to study consciousness by the selfsame instrument of consciousness—the only tool we possess for the purpose.

But there is a vast encouragement in all this, for all of us. We are at our beginning. Human beings are, after all, a spectacularly juvenile species. As evolutionary time is measured, we have only just turned up and have hardly had time to catch our breath, still marveling at our thumbs, still learning to use the brand-new gift of language. Being so young, we can be excused all sorts of folly and can permit ourselves the hope that someday, as a species, we will begin to grow up. What it takes is survival, the hard assignment just ahead in this century. Past that hump, if we can pass it, there is wide-open country ahead.

Science is itself a kind of reassurance that we have the capacity to mature. One of our central, nagging problems is whether we can succeed as the compulsive, obsessive, genetically driven social species that we plainly are. Science is, so far, a small-scale, graspable model which suggests that we really can learn to function as a collective species, in the same sense that ants and termites seem to succeed at their level. The scale is, of course, much smaller for science than for the social insects; there are, I suppose, no more than a million or so genuine scientists in the earth's population. But their behavior is instructive and allusive. The workers are, in the first place, the most individual of individuals, competing with each other in such wild ferocity as to seem, sometimes, almost predatory. Each individual is the most private of selves, telling himself his own private story about nature and then running his private experiment to confirm it or disprove it, then making up another new story and testing again, always feeling himself to be alone and on his own, sometimes looking over his shoulder in apprehension that someone else may be catching up. It looks, near up, like the most solitary kind of human endeavor, matching poetry or art in this respect. But the moment the information comes to hand, it is no longer private; the solitary scientist calls all his friends on the telephone to tell them the news; he sends out preprints and reprints; he turns up at congresses to lecture about it; he shouts from his rooftop.

And what then is made, from all this single-minded work of single minds, is an immense network of knowledge in which all the pieces are fitted together as if by magic. It is a termitarium of information—vaults, arches, nests, and all.

It is constructed, I believe, by something like instinct. I do not wish here to entangle myself in the current wrangle over genes and behavior, except to make four axiomatic assertions about sociobiology: first, we are, biologically speaking, a social species, more so than the ants; second, we are fixed for social living by language, and our capacity to make variable strings of words with syntax is coded in our genes; third, we are, essentially and instinctively, exploring animals;

fourth, we are held together by a separate instinct which dominates all the others, controlling our behavior as individuals; namely, the incessant drive, bred into the bones of each human being, to be *useful.* This last is the hardest and chanciest part of being human, for it requires levels of sensitivity and taste, learnable only by living together, that most of us never reach. Trying to be useful and failing at it is the major source of human discontent, driving some of us crazy.

Science, then, is a model system for collective human behavior and has value because of this for all of us, for it is an activity that can be scrutinized and studied. Quite apart from the quality and interest of the scientific information itself, the phenomenon is intrinsically heuristic and therefore valuable.

For this reason, among others, we have need of critics in science, in the sense that classical art and architecture needed the Ruskins, music the Toveys, and contemporary literature the Leavises, Eliots, and Wilsons. We need people who can tell us how science is done, down to the finest detail, and also why it is done, how the new maps of knowledge are being drawn, and how to distinguish among good science, bad science, and nonsense.

Horace Judson has already staked out his career in this line of work, and the contents of this book are a further stage in his development of his own field. What Judson is up to should not be misconstrued as one more effort to formulate a "philosophy" of science, nor is it in the bailiwick of the history or sociology of science, nor is it scientific journalism. It is criticism, including aesthetic criticism of a high order, based on almost as intimate a knowledge of what goes on in a laboratory as that possessed by the investigator sitting at the bench. I do not know how Judson has succeeded in mastering the stores of information from so many different disciplines, nor how he manages to retain it, nor how he puts it all together in a narrative so engrossing. It is a satisfaction to reflect that he is a young man, with a generation of science still ahead to be scrutinized and weighed, and time enough to set the kind of example needed for the evolution of a school.

June 4, 1979

The Search for Solutions

1

Investigation: The Rage to Know

Certain moments of the mind have a special quality of well-being. A mathematician friend of mine remarked the other day that his daughter, aged eight, had just stumbled without his teaching onto the fact that some numbers are prime numbers—those, like 11 or 19 or 83 or 397, that cannot be divided by any other integer (except, trivially, by 1). "She called them 'unfair' numbers," he said. "And when I asked her why they were unfair, she told me, 'Because there's no way to share them out evenly.'" What delighted him most was not her charming turn of phrase nor her equitable turn of mind (17 peppermints to give to her friends?) but—as a mathematician—the knowledge that the child had experienced a moment of pure scientific perception. She had discovered for herself something of the way things are.

The satisfaction of such a moment at its most intense—and this is what ought to be meant, after all, by the tarnished phrase "the moment of truth"—is not easy to describe. It partakes at once of exhilaration and tranquillity. It is luminously clear. It is beautiful. The clarity of the moment of discovery, the beauty of what in that moment is seen to be true about the world, is the most fundamental attraction that draws scientists on.

Science is enormously disparate—easily the most varied and diverse of human pursuits. The scientific endeavor ranges from the study of animal behavior all the way to particle physics, and from the purest of mathematics back again to the most practical problems of shelter and hunger, sickness and war. Nobody has succeeded in catching all this in one net. And yet the conviction persists—scientists themselves believe, at heart—that behind the diversity lies a unity. In those luminous moments of discovery, in the various approaches and the painful tension required to arrive at them, and then in the community of science, organized worldwide to doubt and criticize, test and exploit discoveries—somewhere in that constellation there are surely constants, to begin with. Deeper is the lure that in the bewildering variety of the world as it is there may be found some astonishing simplicities.

Philosophers, and some of the greatest among them, have offered descriptions of what they claim is the method of science. These make most scientists acutely uncomfortable. The descriptions don't seem to fit what goes on in the doing of science. They seem at once too abstract and too limited. Scientists don't believe that they think in ways that are wildly different from the way most people think at least in some areas of their lives. "We'd be in real trouble—we could get nowhere—if ordinary methods of inference did not apply," Philip Morrison said in a conversation a while ago. (Morrison is a theoretical physicist at the Massachusetts Institute of Technology.) The wild difference, he went on to say, is that scientists apply these everyday methods to areas that most people never think about seriously and carefully. The philosophers' descriptions don't prepare one for either this ordinariness or this extreme diversity of the scientific enterprise—the variety of things to think about, the variety of obstacles and traps to understanding, the variety of approaches to solutions. They hardly acknowledge the fact that a scientist ought often to find himself stretching to the tiptoe of available technique and apparatus, out beyond the frontier of the art, attempting to do something whose difficulty is measured most significantly by the fact that it has never been done before. Science is carried on—this, too, is obvious—in the field, in the observatory, in the laboratory. But historians leave out the arts of the chef and the watchmaker, the development at the bench of a new procedure or a new instrument. "And *making it work,*" Morrison said. "This is terribly important." Indeed, biochemists talk about "the cookbook." Many a Nobel Prize has been awarded not for a discovery, as such, but for a new technique or a new tool that opened up a whole field of discovery. "I am a theoretician," Morrison said. "And yet the most important problem for me is to be in touch with the people who are making new instruments or finding new ways of observing, and to try to get them to do the right experiments." And then, in a burst of annoyance, "I feel very reluctant to give any support to descriptions of 'scientific method.' The scientific enterprise is very difficult to model. You have to look at what scientists of all kinds *actually do.*"

It's true that by contrast philosophers and historians seem book-bound—or, rather, paper-blindered, depending chiefly on what has been published as scientific research for their understanding of the process of discovery. In this century, anyway, published papers are no guide at all to how scientists get the results they report. We have testimony of the highest authenticity for that. Sir Peter Medawar is one who has both done fine science and written well about how it is done: he won his Nobel Prize for investigations of immunological tolerance, which explained, among other things, why foreign tissue, like a kidney or a heart, is rejected by the body into which it is transplanted, and he has described the real methods of science in essays of grace and distinction. A while ago, Medawar wrote, "What scientists *do* has never been the subject of a scientific . . . inquiry. It is no use looking to scientific 'papers', for they not merely conceal but actively misrepresent the reasoning that goes into the work they describe." The observation has become famous, its truth acknowledged by other scientists. Medawar wrote further, "Scientists are building explanatory structures, *telling stories* which are scrupulously tested to see if they are stories about real life."

Scientists do science for a variety of reasons, of course, and most of them are familiar to the sculptor, say, or to the surgeon or the athlete or the builder of bridges: the professional's pride in skill; the swelling gratification that comes with recognition accorded by colleagues and peers; perhaps the competitor's fierce appetite; perhaps ambition for a kind of fame more durable than most. At the beginning is curiosity, and with curiosity the delight in mastery—the joy of figuring it out that is the birthright of every child. I once asked Murray Gell-Mann, a theoretical physicist, how he got started in science. His answer was to point to the summer sky: "When I was a boy, I used to ask all sorts of simple questions—like, 'What holds the clouds up?' " Rosalind Franklin, the crystallographer whose early death deprived her of a share in the Nobel Prize that was given for the discovery of the structure of DNA (the stuff that genes are made of), one day was helping a young collaborator draft an application for research money, when she looked up at him and said, "What we can't tell them is that it's so much *fun!*" He still remembers her glint of mischief. The play of the mind, in an almost childlike innocence, is a pleasure that appears again and again in scientists' reflections on their work. The geneticist Barbara McClintock, as a woman in American science in the 1930s, had no chance at the academic posts open to her male colleagues, but that hardly mattered to her. "I did it because it was *fun!*" she said forty years later. "I couldn't wait to get up in the morning! I never thought of it as 'science.'"

The exuberant innocence can be poignant. François Jacob, who won his share of a Nobel Prize as one of the small group of molecular biologists in the fifties who brought sense and order into the interactions by which bacteria regulate their life processes, recently read an account I had written of that work, and said to me with surprise and an evident pang of regret, "We were like children playing!" He meant the fun of it—but also the simplicity of the problems they had encountered and the innocence of mind they had brought to them. Two hundred and fifty years before—although Jacob did not consciously intend the parallel—Isaac Newton, shortly before his death, said:

> I do not know what I may appear to the world, but to myself I seem to have been only like a boy playing on the sea shore, and diverting myself in now and then finding a smoother pebble or a prettier shell than ordinary, whilst the great ocean of truth lay all undiscovered before me.

For some, curiosity and the delight of putting the world together deepen into a life's passion. Sheldon Glashow, a fundamental-particle physicist at Harvard, also got started in science by asking simple questions. "In eighth grade, we were learning about how the earth goes around the sun, and the moon around the earth, and so on," he said. "And I thought about that, and realized that the Man in the Moon is always looking at us"—that the moon as it circles always turns the same face to the earth. "And I asked the teacher, 'Why is the Man in the Moon always looking at us?' She was pleased with the question—but said it was hard. And it turns out that it's not until you're in college-level physics courses that one really learns the answers," Glashow said. "But the *difference* is, that most people would look at the moon and wonder for a moment, and say, 'That's interesting'—and then forget it. But some people can't let go."

Curiosity is not enough. The word is too mild by far, a word for infants. Passion is indispensable for creation, no less in the sciences than in the arts. Medawar once described it in a talk addressed to young scientists. "You must feel in yourself an exploratory impulsion—an *acute discomfort* at incomprehension." This is the rage to know. The other side of the fun of science, as of art, is pain. A problem worth solving will surely require weeks and months of lack of progress, whipsawn between hope and the blackest sense of despair. The marathon runner or the young swimmer who would be a champion knows at least that the pain may be a symptom of progress. But here the artist and the scientist part company with the athlete—to join the mystic for a while. The pain of creation, though not of the body, is in one way worse. It must not only be endured but reflected back on itself to increase the agility, variety, inventiveness of the play of the mind. Some problems in science have demanded such devotion, such willingness to bear repeated rebuffs, not just for years but for decades. There are times in the practice of the arts, we're told, of abysmal self-doubt. There are like passages in the doing of science. Albert Einstein took eleven years of unremitting concentration to produce the general theory of relativity; long afterward, he wrote, "In the light of knowledge attained, the happy achievement seems almost a matter of course, and any intelligent student can grasp it without too much trouble. But the years of anxious searching in the dark, with their intense longing, their alternations of confidence and exhaustion, and the final emergence into the light—only those who have experienced it can understand it." Einstein confronting Einstein's problems: the achievement, to be sure, is matched only by Newton's and perhaps Darwin's—but the experience is not rare. It is all but inseparable from high accomplishment. In the black cave of unknowing, when one is groping for the contours of the rock and the slope of the floor, tossing a pebble and listening for its fall, brushing away false clues as insistent as cobwebs, a touch of fresh air on the cheek can make hope leap up, an unexpected scurrying whisper can induce the mood of the brink of terror. "Afterwards it can be told—trivialized—like a *roman policier*, a detective story," François Jacob once said. "While you're there, it is the sound and the fury." But it was the poet and adept of mysticism St. John of the Cross who gave to this passionate wrestling with bafflement the name by which, ever since, it has been known: *the dark night of the soul.*

Enlightenment may not appear, or not in time; the mystic at least need not fear forestalling. Enlightenment may dawn in ways as varied as the individual approaches of scientists at work—and, in defiance of stereotypes, the sciences far outrun the arts in variety of personal styles and in the crucial influence of style on the creative process. During a conversation with a co-worker—and he just as baffled—a fact quietly shifts from the insignificant background to the foreground; a trivial anomaly becomes a central piece of evidence, the entire pattern swims into focus, and at last one sees. "How obvious! We knew it all along!" Or a rival may publish first but yet be wrong—and in the crashing wave of fear that he's got it right, followed and engulfed by the wave of realization that it must be wrong, the whole view of the problem skews, the tension of one's concentration twists abruptly higher, and at last one sees. "Not that way, *this* way!"

One path to enlightenment, though, has been reported so widely, by writers and artists, by scientists, and especially by mathematicians, that it has become established as a discipline for courting inspiration. The first stage, the reports agree, is prolonged contemplation of the problem, days of saturation in the data, weeks of incessant struggle—the torment of the unknown. The aim is to set in motion the unconscious processes of the mind, to prepare the intuitive leap. William Lipscomb, a physical chemist at Harvard who won a Nobel Prize for finding the unexpected structures of some unusual molecules, the boranes, said recently that, for him, "The unconscious mind pieces together random impressions into a continuous story. If I really want to work on a problem, I do a good deal of the work at night—because then I worry about it as I go to sleep." The worry must be about the problem intensely and exclusively. Thought must be free of distraction or competing anxieties. Identification with the problem grows so intimate that the scientist has the experience of the detective who begins to think like the terrorist, of the hunter who feels, as though directly, the silken ripple of the tiger's instincts. One great physical chemist was credited by his peers, who watched him awestruck, with the ability to think about chemical structures directly in quantum terms—so that if a proposed molecular model was too tightly packed he felt uncomfortable, as though his shoes pinched. Joshua Lederberg, president of the Rockefeller University, who won his Nobel for discoveries that established the genetics of microorganisms, said recently, "One needs the ability to strip to the essential attributes of some actor in a process, the ability to imagine oneself *inside* a biological situation; I literally had to be able to think, for example, 'What would it be like if I were one of the chemical pieces in a bacterial chromosome?'—and to try to understand what my environment was, try to know *where* I was, try to know when I was supposed to function in a certain way, and so forth." Total preoccupation to the point of absentmindedness is no eccentricity—just as the monstrous egoism and contentiousness of some scientists, like some artists, are the overflow of the strength and reserves of sureness they must find how they can.

Sometimes out of that saturation the answer arises, spontaneous and entire, as though of its own volition. In a famous story, Friedrich Kekulé, who was a German chemist of the mid-nineteenth century, described how a series of discoveries came to him in the course of hypnagogic reveries—waking dreams. His account, though far from typical, is supremely charming. Kekulé was immersed in one of the most perplexing problems of his day: to find the structural basis of organic chemistry—that is, of the chemistry of compounds that contain carbon atoms. Enormous numbers of such compounds were coming to be known, but their makeup—from atoms of carbon, hydrogen, oxygen, and a few other elements—seemed to follow no rules. Kekulé had dwelt on the compounds' behavior so intensely that the atoms themselves on occasion seemed to appear to him and dance. In the dusk of a summer evening, he was going home by horse-drawn omnibus, sitting outside and alone. "I fell into a reverie, and lo! The atoms were gamboling before my eyes," he later wrote. "I saw how, frequently, two smaller atoms united to form a pair; how a larger one embraced two smaller ones; how still larger ones kept hold of three or even four of the smaller; whilst the whole kept whirling in a giddy dance. I saw how the larger ones formed a chain." He spent hours that night sketching the forms he had envisioned. Another time, when Kekulé was nodding in his chair before the fire, the atoms danced for him again—but only the larger ones, this time, in long rows, "all twining and twisting in snakelike motion. But look! What was that? One of the snakes had seized hold of its own tail, and the form whirled mockingly before my eyes." The chains and rings that carbon atoms form with each other are indeed the fundamental structures of organic chemistry.

Although without Kekulé's vivid details, several scientists have told me that the fringes of sleep set the problem-sodden mind free to make uninhibited, bizarre, even random connections that may throw up the unexpected answer. One said that the technical trick that led to one of his most admired discoveries—it was about the fundamental, molecular nature of genetic mutations—had sprung to mind while he was lying insomniac at three in the morning. Another said he was startled from a deep sleep one night by the fully worked-out answer to a puzzle that had blocked him for weeks—except that at breakfast he was no longer able to remember any detail except the jubilant certainty. So the next night he went to sleep with paper and pencil on the bedside table; and when, once again, he awoke with the answer he was able to seize it.

More usually, though, in the classic strategy for achieving enlightenment the weeks of saturation must be followed by a second stage that begins when the problem is deliberately set aside—put out of the active mind, the ceaseless pondering switched off. After several days of silence, the solution wells up. The mathematician Henri Poincaré was unusually introspective about the process of discovery. (He also came nearer than anyone else to beating Einstein to the theory of relativity, except that in that case, though he had the pieces of the problem, inspiration did not strike.) In 1908, Poincaré gave a lecture, before the Psychological Society of Paris, about the psychology of mathematical invention, and there he described how he made some of his youthful discoveries. He reassured his audience, few of them mathematical: "I will tell you that I found the proof of a certain theorem in certain circumstances. The theorem will have a barbarous name, which many of you will never have heard of. But that's of no importance, for what is interesting to the psychologist is not the theorem—it's the circumstances." The youthful discovery was about a class of mathematical functions which he named in honor of another mathematician, Lazarus Fuchs—but, as he said, the mathematical content is not important here. The young Poincaré believed, and for fifteen days he strove to prove, that no functions of the type he was pondering could exist in mathematics. He struggled with the disproof for hours every day. One evening, he happened to drink some black coffee, and couldn't sleep. Like Kekulé with his carbon atoms, Poincaré found mathematical expressions arising before him in crowds, combining and recombining. By the next morning, he had established a class of the functions that he had begun by denying. Then, a short time later, he left town to go on a geological excursion for several days. "The changes of travel made me forget my mathematical work." One day during the excursion, though, he was carrying on a conversation as he was about to board a bus. "At the moment when I put my foot on the step, the idea came to me, without anything in my former thoughts seeming to have paved the way for it, that the transformations I had used to define the Fuchsian functions were identical with those of non-Euclidian geometry." He did not try to prove the idea, but went right on with his conversation. "But I felt a perfect certainty," he wrote. When he got home, "for conscience's sake I verified the result at my leisure."

The quality of such moments of the mind has not often been described successfully; Charles P. Snow was a scientist and a novelist as well, and when his experience of science came together with his writer's imagination his witness is assured and authentic. In *The Search*, a novel about scientists at work, the protagonist makes a discovery for which he had long been striving.

> Then I was carried beyond pleasure. . . . My own triumph and delight and success were there, but they seemed insignificant beside this tranquil ecstasy. It was as though I had looked for a truth outside myself, and finding it had become for a moment a part of the truth I sought; as though all the world, the atoms and the stars, were wonderfully clear and close to me, and I to them, so that we were part of a lucidity more tremendous than any mystery.
>
> I had never known that such a moment could exist. . . . Since then I have never quite regained it. But one effect will stay with me as long as I live; once, when I was young, I used to sneer at the mystics who have described the experience of being at one with God and part of the unity of things. After that afternoon, I did not want to laugh again; for though I should have interpreted the experience differently, I thought I knew what they meant.

This experience beyond pleasure, like the dark night of the soul, has a name: the novelist Romain Rolland, in a letter to Sigmund Freud, called it "the oceanic sense of well-being."

Science is our century's art. Nearly four hundred years ago, when modern science was just beginning, Francis Bacon wrote that *knowledge is power.* Yet Bacon was not a scientist. His slogan was the first clear statement of the promise by which, ever since, bureaucrats justify to each other and to king or taxpayer the spending of money on science. Knowledge is power; today we would say, less grandly, that science is essential to technology. Bacon's promise has been fulfilled abundantly, magnificently. The rage to know has been matched by the rage to make. Therefore—with the proviso, abundantly demonstrated, that it's rarely possible to predict which program of fundamental research will produce just what technology and when—the promise has brought scientists in the Western world unprecedented freedom of inquiry. Nonetheless, Bacon's promise hardly penetrates to the thing that moves most scientists. Science has several rewards, but the greatest is that it is the most interesting, difficult, pitiless, exciting, and beautiful pursuit that we have yet found. Science is our century's art.

The takeover can be dated more precisely than the beginning of most eras: Friday, June 30, 1905, will do, when Albert Einstein, a clerk in the Swiss patent office in Bern, submitted a thirty-one-page paper, "On the Electrodynamics of Moving Bodies," to the journal *Annalen der Physik.* No poem, no play, no piece of music written since then comes near the theory of relativity in its power, as one strains to apprehend it, to make the mind tremble with delight. Whereas fifty years ago it was often said that hardly twoscore people understood the theory of relativity, today its essential vision, as Einstein himself said, is within reach of any reasonably bright high school student—and that, too, is characteristic of the speed of assimilation of the new in the arts.

Consider also the molecular structure of that stuff of the gene, the celebrated double helix of deoxyribonucleic acid. This is two repetitive strands, one winding up, the other down, but hooked together, across the tube of space between them, by a sequence of pairs of chemical entities—just four sorts of these entities, making just two kinds of pairs, with exactly ten pairs to a full turn of the helix. It's a piece of sculpture. But observe how form and function are one. That sequence possesses a unique duality: one way, it allows the strands to part and each to assemble on itself, by the pairing rules, a duplicate of the complementary strand; the other way, the sequence enciphers, in a four-letter alphabet, the entire specification for the substance of the organism. The structure thus encompasses both heredity and embryological growth, the passing-on of potential and its expression. The structure's elucidation, in March of 1953, was an event of such surpassing explanatory power that it will reverberate through whatever time mankind has remaining. The structure is also perfectly economical and splendidly elegant. There is no sculpture made in this century that is so entrancing.

If to compare science to art seems—in the last quarter of this century—to undervalue what science does, that must be, at least partly, because we now expect art to do so little. Before our century, everyone of course supposed that the artist imitates nature. Aristotle had said so; the idea was obvious, it had flourished and evolved for two thousand years; those who thought about it added that the artist imitated not just nature as it accidentally happens, but by penetrating to nature as it has to be. Yet today that describes the scientist. "Scientific reasoning," Medawar also said, "is a constant interplay or interaction between hypotheses and the logical expectations they give rise to: there is a restless to-and-fro motion of thought, the formulation and reformulation of hypotheses, until we arrive at a hypothesis which, to the best of our prevailing knowledge, will satisfactorily meet the case." Thus far, change only the term "hypothesis" and Medawar described well the experience the painter or the poet has of his own work. "Scientific reasoning is a kind of dialogue between the possible and the actual, between what might be and what is in fact the case," he went on—and there the difference lies. The scientist enjoys the harsher discipline of what is and is not the case. It is he, rather than the painter or the poet in this century, who pursues in its stringent form the imitation of nature.

Many scientists—mathematicians and physicists especially—hold that beauty in a theory is itself almost a form of proof. They speak, for example, of "elegance." Paul Dirac predicted the existence of antimatter (what would science fiction be without him?) several years before any form of it was observed. He won a share in the Nobel Prize in physics in 1933 for the work that included that prediction. "It is more important to have beauty in one's equations than to have them fit experiment," Dirac wrote many years later. "It seems that if one is working from the point of view of getting beauty in one's equations, and if one has really a sound insight, one is on a sure line of progress."

Here the scientist parts company with the artist. The insight must be sound. The dialogue is between what might be and what is in fact the case. The scientist is trying to get the thing right. The world is there.

And so are other scientists. The social system of science begins with the apprenticeship of the graduate student with a group of his peers and elders in the laboratory of a senior scientist; it continues to collaboration at the bench or the blackboard, and on to formal publication—which is a formal invitation to criticism. The most fundamental function of the social system of science is to enlarge the interplay between imagination and judgment from a private into a public activity. The oceanic feeling of well-being, the true touchstone of the artist, is for the scientist, even the most fortunate and gifted, only the midpoint of the process of doing science.

A Conversation with Murray Gell-Mann

The inventor of quarks talks about how, in physics, beauty points the way to truth.

Most physicists have a profound conviction that nature is inherently orderly and that the order can be discovered. Murray Gell-Mann is a theoretical physicist at the California Institute of Technology. He won a share of the Nobel Prize in physics in 1969 for his work on subatomic particles. In 1963, he first conceived the entities that may, after all, be the stuff that all other stuff is made of—and he named them "quarks." The word appears in James Joyce's *Finnegans Wake*, in the puzzling phrase "Three quarks for muster mark," and it's true that in Gell-Mann's original proposal it took three quarks to make a proton. But he and other strange-particle physicists pronounce it to rhyme with "fork" rather than "lark." The bizarre term seems to stick in everybody's mind. Quarks and related fundamental entities obey forces, besides the everyday gravitational and electromagnetic fields, that are known only within the diameter of the atomic nucleus—almost inconceivably short—and that are called the strong interaction force and the weak. They display properties, different from such homely ideas as mass, charge, or spin, that can only be described mathematically, we're told, but that have been named "strangeness," "charm," "color," and "flavor." Gell-Mann spends his summers in the Rocky Mountains at Aspen, Colorado, where during the days he talks with other physicists from all over the world who gather at the Aspen Physics Center and in the evenings goes with his wife, Margaret, an archaeologist, to concerts at the Aspen Music Tent. After a concert, one night, we all went to dinner at Arthur's, Aspen's Chinese restaurant. When Gell-Mann had ordered for everybody in Chinese—he has learned Chinese, he explained, the way someone else might learn menu French, just enough to be sure of getting what he wants in Chinese restaurants anywhere—he talked about science.

I asked, where do ideas come from?

"Everybody agrees on that," Gell-Mann said. "We had a seminar here, about ten years ago, including several painters, a poet, a couple of writers, and the physicists. Everybody agrees on how it works. All of these people, whether they're doing artistic work or scientific work, are trying to solve a problem.

"Any art that's worth the name has some kind of discipline associated with it," he went on. "Some kind of a rule—maybe it's not the rule of a sonnet, or a symphony, or a classical painting, but even the most *liberated* contemporary art, if it's any kind of art at all, has some kind of rule. And the object is to get across what you're trying to get across, while sticking to the rules.

"In *our* business, it's hard to find out what nature is up to, and there are all sorts of constraints—you have to agree with everything you already know, you have to have a nice self-consistent structure, and so on. And so, when you have to account for something new—or in the artists' case, when you want to express yourself within those rules—you run into *problems.* And the problems at first seem difficult, and perhaps insoluble. And you work very hard trying to understand, trying to *fill yourself full* of the problem, just to know what barriers you're trying to crack.

"And after that, further voluntary effort, further conscious effort, is not so productive. And at that point, what the shrinks would call the preconscious, I guess, seems to be more important. Processes that are just outside of awareness go on, which thrust up bubbles of ideas from time to time. And that can happen when you're driving, or shaving, or walking—anything.

"One of my ideas came in a slip of the tongue. I was getting up at a seminar—one person had just put forward a theory, and I was explaining why his theory was wrong, why it didn't work—and while I was explaining it, I happened to blurt out the correct way to do it. Just a slip of the tongue. And I recognized immediately that that would solve the problem."

What was that?

"Oh, it was the idea of 'strangeness,'" Gell-Mann said. "I was *intending* to say, 'isotopic spin—three halves. And, instead, I said, 'isotopic spin—*one.*' Which, for the particular kind of particle, was *unthinkable.* Unthinkable—but correct. Everybody had taken it for granted that the value was a half integer. And it wasn't. And as soon as I mentioned an integral number, which nobody had thought of, I realized that the integral number was the answer."

What had led him to the name "quarks"?

"It was a nice sound," Gell-Mann said. "And then I found it in *Finnegans Wake.*"

And those terms "flavor" and "color" and "charm"?

"Oh, charm is a flavor. Just one of the flavors. There are many flavors. We used to know three—up, down, and strange. But now there's charm, and there's several more too, probably. The terms? Just for fun. I was one of those who introduced 'color,' I think. And Yoichura Nambu at the University of Chicago introduced 'flavor.' There's no particular reason to give pompous names, one might as well give playful names—it doesn't matter. But the material is very serious—it really works beautifully."

Why do people work on science?

"Well, I think most of the clichés are literally true—people are driven by curiosity," he said.

"Another thing is, I suspect, that science is more tempting to people who, when they're young, don't very much enjoy the interplay of human beings—which occurs in business, for example. They'd rather interact with nature than do something that depends exclusively on interacting with other people. Of course, there are some scientists who play it against other people—but most scientists think that nature is the opponent."

I observed that the physicist Paul Dirac once said that beauty, elegance in a theory, is almost what matters most.

"Yes—though we don't know what that means," Gell-Mann said. "There's a quotation from Newton, I don't remember the exact words but lots of other physicists have made the same remark since—that nature seems to have a remarkable property of self-similarity. The laws—the fundamental laws—at different levels seem to resemble one another. And that's probably what accounts for the possibility of using elegance as a criterion. We develop a mathematical formula, say, for describing something at a particular level, and then when we go to a deeper level we find that, in terms of mathematics, the equations at the deeper level are beautifully equivalent. Which means that we've found an appropriate formula.

"And that takes the human being, human judgment, out of it a little. You might object that after all *we* are the ones who say what elegance is. But I don't think that that's the point. One way to describe what's going on is to say that nature apparently *resembles itself* at different levels."

But how much did he find himself consciously applying a criterion of elegance?

"Constantly! Nothing but. Well, I mean, that together with trying to get the thing right! Trying to explain the data. But, you know, frequently a theorist will even *throw out* a lot of the data on the grounds that if they don't fit an elegant scheme, they're wrong. That's happened to me many times."

For example?

"Oh, the theory of the weak interaction. There were *nine* experiments that contradicted it—all wrong. Every one.

"When you have something simple that agrees with all the rest of physics and really seems to explain what's going on, a few experimental data against it are no objection whatever. Almost certain to be wrong. Einstein was not very much perturbed when some early experiments showed the special theory of relativity to be wrong."

I had heard that certain criteria of symmetry seem to determine the number and types and relations of fundamental particles. Was this so?

"Well, a symmetry that is *almost* there and then, in the results, is broken," Gell-Mann said. "The *equations* appear to be symmetrical. But nature chooses an unsymmetrical set of solutions. Something very subtle. The violations of symmetry among different kinds of quarks apparently have that characteristic. And yet, *some* symmetries seem to be perfect—like the 'color' symmetry of quarks."

Almost as if the symmetry of part of the system was itself an asymmetry in the entire system?

"But there is some specific mathematical mechanism by which all this comes about, which we're just beginning to understand—very dimly," he said. "Everything we do is explained mathematically. But there's apparently some beautifully subtle way in which perfectly symmetrical equations produce asymmetrical physics, in certain cases."

The supersystem, in order to be *perfectly* not-quite-perfect, would have to have the imperfection of being perfectly symmetrical in some places?

"That's exactly, apparently, what happens. Yes. And we don't understand it perfectly. We're beginning to understand it."

Murray Gell-Mann receiving the Nobel Prize in Physics from King Gustav VI Adolf of Sweden, in Stockholm, December 10, 1969.

2

Pattern

Arriving in London again a while ago, after a time away, my wife and I came to the door of the plane and looked out. It was seven-thirty of a summer's morning there, half-past two in the night back in New York. After the long tunnel of the flight, we were tired yet keyed up, and slightly disoriented. We started along the ramp and up the corridor toward passport and customs clearance. And the tiny insistent differences in the rhythms of life began to bear in.

"Business trip? Holiday?" As the immigration officer stamps my passport, somewhere a telephone is ringing—not the steady long *beeeeep,* pause, *beeeeep,* pause of an American phone, but the short English *beep-beep* pause *beep-beep.* The bell's note is shriller, too. I have no problem understanding the bank clerk when I change a traveler's check, but twice I have to ask the porter with our bags to repeat himself, and in turn the taxi driver has trouble catching our destination. I've been here before; it's not the choice of words nor even their pronunciation that gets in the way, but the rhythms of quickly spoken sentences, different enough to upset expectations. As the taxi reaches the city, the noises in the street are pitched much higher. Smaller cars here, and their smaller engines turn over faster and so sing a higher note. Trucks are smaller, too, and don't roar so much as whine. The sound of traffic outside our hotel is not so loud as in the States, yet insistent, hectic, the bass notes missing.

In the room, the phone is nagging *beep-beep.* An invitation for theater and dinner tonight—to have them in that sequence. My wife tries to take a morning nap. Our personal rhythms of sleeping and waking, eating and digestion, the rise and fall of body temperature and with that our alertness, will require several days to come around to the five-hour time change—but this, the shift in circadian rhythms otherwise called jet lag, we know about as most travelers do these days, and are prepared to suffer through. Jet lag is only the most obvious change in the patterns and cycles around us. Tea at four-thirty makes four meals a day and drives dinner later. My wife's hair dryer is wired with a conversion switch so it won't burn out on the European 240 volts—but she notices that its fan is weaker. And it hums a deeper tone. It's driven by current at 50 cycles per second rather than 60. In London, the theater curtain goes up at seven-thirty. Late that evening, on the way to dinner after the play, the sky is still bright. London is far more northerly than any place in the forty-eight contiguous states. It's as far north as parts of Hudson's Bay, and the midsummer nights are hours shorter than in New York. Even the rhythm of the seasons is different: the celebrated English springtime really does begin about the twenty-first of March and summer weather arrives, if at all, late in June.

Beat of the traffic, pulse of the phone, the long cycles of the angle of the sun in the sky. Patterns, rhythms. We live by patterns. Intervals. Repetitions. Patterns set up expectations. Patterns in time. To perceive a pattern means that we have already formed an idea what's next. Rhythms in space. A great scientist said that there's no science without measurement and quantity—but he was wrong, for in science as in life, patterns come even before numbers. Natural patterns, man-made patterns. Home for the hunter is a pattern of two hills and a clump of pines—and an expectation. The spiral of a snail's shell, the spiral of the great nebula in Andromeda. The punch line of a joke tells us that a set of things we thought belonged to one pattern was really, all along, making a very different pattern. Folding of rocks and meanders of rivers. Music is *all* pattern—too regular and the music is banal, but a great composer teases our sense of pattern, upsets expectations but then resolves the complexity by reimposing the pattern at a more encompassing level. Symmetry. Broken symmetry. Mathematics itself, in large part, is the recognition and pursuit of patterns in numbers.

Patterns make connections. A subatomic event is read in a pattern of curved or straight tracks in a photographic emulsion—and here, to make a discovery is to encounter a break from the expected pattern, for which the theorist must then find pattern again at a more encompassing level. But can we also say that to make great music is to play a game of theory-building with the listener? We live by patterns. The individuality of a loved face resides precisely in its broken symmetry—the two halves not quite alike. Patterns tell us where we are.

Man in ten thousand years has superimposed his own patterns and rhythms on the world, and ours are often more complex than those that would persist without us. But millennia before that began, the growth and diversity and ubiquitous spread of living beings vastly elaborated the patterns of preanimate earth. The world we are transforming was already transformed by life. There's more syncopation for the eye and mind in a square mile of Alpine meadow or Sumatran rain forest than in all of Mars and Jupiter with our moon thrown in.

Put matter into space, and the ways it can pack, or bend, or bulge, or flow, or break, or splash are stringently and inescapably controlled by the facts of geometry and of energy. The patterns of nature are built from a very few themes. Economies prevail. The world so abundant is at the same time parsimonious. Things make patterns, insofar as they do so at all, because they cannot do otherwise.

Thus, as perhaps the simplest example, certain angles appear again and again. Drill a hole in a granite boulder, pour it full of water, freeze it until the expanding plug cracks the rock. The most likely break will make three chunks, and the cracks will intersect at an angle as close to 120° as the local variations in the grain of the rock will allow. The earth around a houseplant, or a flat sheet of mud after a rainstorm—as they dry and shrink, they crack, and the cracks outline irregular cells, yet come together almost always at triple joints with angles as close as may be to 120°. On an antique Chinese vase of rare delicacy, the filigree of crazing beneath the topmost layer of transparent glaze makes a pattern of cracks and angles fundamentally similar to the dried mud, and for the same reasons. Rock, mud, glaze, the cracking relieves strains; the patterns of pieces must be similar and the angles repetitious because these relieve the strains most efficiently. Soap bubbles—Peter Stevens, an architect, points this out in a clever, handsome book, *Patterns in Nature*—provide a model for all the films, membranes, and elastic surfaces occurring so widely. A pair or a cluster of soap bubbles is pushed by the balancing of pressures and sizes—pushed by economy of energy and geometry—into a momentarily stable configuration in which, as photographs record permanently, once more the angles of meeting are 120°.

Constraints like these rule the forms created in life, too. Stevens shows a pair of pictures, one of a cluster of bubbles against a black ground, the other of the plates in the armor of a box turtle—a startling near-identity of pattern. Throughout nature, inanimate or animate, one finds that spherical objects or rodlike ones, whether molecules, viruses, grains, cells, pack with their own kind in the closest fit, and so in hexagonal arrays—where, of course, the intersections form 120° angles. The hexagonal shapes of the cells of the honeycomb are only the most celebrated example: formed by the serene cooperation of the inherited behavior of the bees with the physical forces that determine the most balanced placement of the wax partitions. Any worker bee from the hive can come to the growing edge of the comb and begin immediately to build on at the correct angle.

The other ubiquitous angle is imposed by the three-dimensional geometry of the simplest pyramid—the pyramid formed of four equilateral triangles, one at the bottom and three sides. This pyramid has four angles, and so is called, from the Greek, the tetrahedron. Tetrahedra also pack to fill space completely. Anything that points four ways in space from a common center tends to point at the four corners of a tetrahedron. The familiar object is the photographer's tripod, if set up with its three legs making equal angles with each other and the central shaft. Those equal angles out from the center are curious: 109° 28′ 16″. The tetrahedral angle is supremely important in modern biology. Its importance is due to the geometry of the carbon atom. The carbon atom is a fundamental and much-multiplied building block of biological molecules. A carbon atom forms chemical bonds with as many as four other atoms—and these bonds splay out from each other as far as they can get, pointing to the four corners of a tetrahedron and thus forming with each other the tetrahedral angle, 109° 28′ 16″.

The recognition of rudimentary patterns in nature has sometimes had profound consequences. Place a sheet of pastry into a pie plate: where it rises from the flat bottom to the rim, the excess dough folds and bulges and must be patted into an even thickness. Fill the pie with apple slices, then join the top crust to the bottom at the edge and seal them. Bake the pie, cut a slice, look at the layers and folds of the crust. The rocks that immediately underlie more than three-quarters of the earth's surface—shales and sandstones, slates and marbles—display layer forms, often upended and deeply folded and twisted, that remind one irresistibly of a giant pastry. We have all seen these, say from the window of the car along a new road where it cuts through a hill. When the geologists of the eighteenth and early nineteenth centuries began to look seriously at these formations—the great canal network of England was then being cut through the crust of the earth—the analogy was obvious that such rocks had been laid down by sediments, then pressed and baked at enormous pressures and temperatures. Where could that oven be except deep in the earth? But such sinkings and risings must have taken time—time on a scale far greater than had ever been imagined, at least in the West. Until then, educated men knew little reason not to accept the biblical arguments of Bishop James Ussher that the earth began in 4004 B.C. The conviction that the earth had a history to be measured in millions, indeed billions, of years grew from many kinds of evidence. None of the evidence was more powerfully persuasive than the visible patterns that a child who has rolled out scraps of his mother's pie dough will recognize in the rocks. And that recognition swings open an enormous door in the mind—to the idea of the ancientness of the world. The conviction that the world had a history opened the way for the idea that the forms of life have a history—the idea of evolution. Darwin acknowledged his greatest intellectual debt to the geologists.

"Meandering"—a synonym for lazy randomness. The distinctive pattern of sinuous curves of the channel of a slow-moving river has been recognized since the ancients; the word itself comes from a river in Asia Minor known to the Greeks as Maiandros. The first satisfactory attempt to explain meanders was made in 1926 by Albert Einstein. Writing about the mechanics of flowing liquids and of turbulence, Einstein observed that even a small curve in a river would cause the water, sweeping past it, to be flung by centrifugal force against the concave farther bank. The water at the top surface of the river is slowed least by friction against the riverbed, and so the top of the current slides faster in the centrifugal swing. As this layer reaches the far bank, its relative force drives it against the bank—scouring as it moves. This layer then has nowhere to go but down, creating a countercurrent along the bottom, from far bank to near. By this swing and roll, the current eats the small curve deeper. The material removed is deposited farther downstream as the river slows and begins to swing in the other direction to keep going downhill—and then scours this next curve the same way.

Surprisingly, Einstein did not penetrate from the mechanical question to the energy. But in the 1950s and '60s, Luna Leopold and colleagues at the United States Geological Survey demonstrated that meanders are the form in which a river distributes most evenly the work it must do in turning and descending. A meander digs here and deposits there in a constant process of adjustment that smooths out the rate at which the river drops. A flexible strip of steel, held at two points and bent hard, takes on shapes that distribute the twisting energy most evenly along the strip. The shape of the curve exactly resembles river meanders. The mathematics of the forms are the same. "Meandering"—lazy, yes, in the precise sense of spreading out the work to eliminate peaks of effort. But far from accidental.

Put matter into space, give it life, make it grow. The diversity of living things is orchestrated, in part, as variations on a very few new themes. Of all these, the most universal are spirals and helices.

A spider spinning her web, after stringing out the frame of spokes, fills in with a spiral of the simplest kind. As she goes around and around, she guides herself by the previous turn—as though holding a handrail—and so the distance between the turns (once she's made the start) is everywhere the same. This spiral appears in the round hooked rug, in the coiled steel tape measure. Its geometry was known to the ancients. It is called the spiral of Archimedes. A spiral of Archimedes, if a straight line is drawn from its origin outward, intersects that line always at the same interval from one whorl to the next. A look at the pattern reveals a geometric peculiarity: because this distance between one intersection and the next stays the same, the angle of the spiral's intersection with the radial line slowly changes, tending more and more toward a right angle.

The spiral most characteristic of the forms of living things ought to be called (though it isn't) the Cartesian spiral, because it was first identified by René Descartes. In a letter to a friend in 1638, Descartes constructed a curve growing outward that cut every radial line always at the same, unchanging angle. This was an analogy with the way a circle meets *its* radius, except that a circle doesn't grow. He then showed that for such a spiral, the distances from whorl to whorl, at which it intersects any radius, though not identical are yet in constant proportion, each to the next.

Descartes's equiangular spiral is one of the most magical curves in geometry. Proportionality pervades it. Proportionality explains its great beauty and accounts for its extraordinary properties. Most deeply, if two lengths of the spiral are cut off by radial lines that make the same angle at the origin, the lengths are similar in every way. From this follows the spiral's most pleasing aspect: it grows continuously yet never changes its shape.

Descartes's equiangular spiral is implicit in many other patterns—as in the piling up of congruent triangles, or the packing of squares or hexagons. Above all, it is the spiral of growth. Almost wherever a creature traces lastingly the continuing stages of its growth, there the equiangular spiral is found. This is the spiral of shells, which are a permanent record of their inhabitants' growth. It is also the form of tusks, horns, and teeth. Whether snail or elephant or beaver, as the creature itself grows, it adds to its shell or tusk or tooth at one end and by progressively greater amounts. That could produce a simple cone—but if the material added is more abundant, even by little, on one side than another, the equiangular spiral must result. The beaver's cutting teeth wear down too fast to display the curve more than slightly, but it is there. The water buffalo has horns triangular at the base (as do many sheep), so that the inequality of growth that adds more at the inside front combines with the pull of the weight of the enlarging horn to produce a compound, three-dimensional version of the spiral.

The greatest student of the relations of growth in living things was D'Arcy Wentworth Thompson, a British natural philosopher—classicist, mathematician, naturalist—who died in 1948, aged eighty-eight. He was fascinated by the multiplicity of patterns in shells and horns that this single geometrical form displays. "The shell, like the creature within it, grows but does not change its shape," he wrote. "The shell retains its unchanging form in spite of its asymmetrical growth; it grows at one end only, and so does the horn. And this remarkable property of increasing by terminal growth, but nevertheless retaining unchanged the form of the entire figure, is characteristic of the equiangular spiral, and of no other mathematical curve." From this inescapable geometry follow the shell of the clam as well as of the triton, the flare of the tusk of the woolly mammoth and the smile on the face of the saber-toothed tiger, the compound elegance of the horn of the bighorn sheep, the still perfection of the chambered nautilus—music to the eye.

The curve still more fundamental to all life, though, is neither so beautiful nor so obviously intriguing. (D'Arcy Thompson knew but almost ignored it.) This is the spiral's simpler relative, the helix—as in the twist of a hawser, the thread of a machine screw. Unlike the spiral, the helix doesn't swing steadily wider from a point of origin. Monotonously, the helix just strings along always at the same diameter. That turns out to be the secret of its importance. The helix is oddly rare in inorganic nature. In living things, it appears infrequently and fleetingly in the forms we see—perhaps in a crisp curl of hair or in the grasping tendril of a vine. The domain of the helix is among the processes of life at the smallest level where we can speak of such processes at all—at the level of the individual molecules that determine all that goes on within the living cell.

The molecules most characteristic of life are very large—far larger than molecules not produced by living processes. (Their only rivals are artificial fibers and certain plastics—and these also, being man-made, are in a perfectly serious sense products of life.) Cells have lots of small molecules in them, too, of course; the large molecules use the small ones for energy and raw materials to manufacture more of themselves. The instructions for all this are carried by one kind of large molecule, the nucleic acids. In particular, deoxyribonucleic acid—DNA—makes up the genes, the blueprint for the organism. The instructions are then carried out by another kind of large molecule, the proteins. In particular, the enzymes are protein molecules that act as the machine tools of the cell. The large molecules are built as long chains of subunits. Nucleic acids are made of hundreds or often thousands of just four sorts of subunits, strung together. Proteins are more varied, being made up from twenty different sorts of subunits, the amino acids, hooked together like beads of slightly different shapes, in a sequence that is exactly the same for every molecule of any particular protein. The great mystery of biology in the 1950s was how the instructions on the nucleic acids get translated into sequences of amino acids in proteins, then to do the work of the organism. But that mystery could not even be thought about seriously until people recognized the patterns on which these long molecules are built.

Linus Pauling, the greatest physical chemist of the century, perceived in 1948 the sort of very simple point on which great reputations in science are often based. He saw that in a sequence of virtually identical objects in space—whatever they are, rosebuds, say, or amino acids—as one follows from the first to the second, the change in position is a combination of movement that must, if carried on, generate a helix. "The idea I had in 1948 is this," he said to me. "You can show, mathematically, that if you have a structure, such as a right hand"—he held his right hand in the air—"and a second structure that is identical with it, but in a different position in space"—he raised his hand higher, turning it—"the relationship between this structure and *this* one"—he rotated his hand up and down several times—"is *to rotate around* an axis and *translate along* the axis. That's the general relation between two asymmetric but equivalent objects in space. And if you repeat this operation, you get a large number of equivalent objects that automatically form a helix." Pauling built, in 1951, a particular helical model of amino acids—the alpha helix—that proved to be frequently found in the structure of proteins. Then in 1953, James Watson and Francis Crick, shaping their approach on Pauling's, found the structure of the genetic material itself, the celebrated double helix of DNA.

Patterns run throughout that story. Watson and Crick were able to beat their rivals to the structure of the double helix because Watson saw a telltale pattern in an X-ray photo of DNA taken by a scientist in another laboratory, Rosalind Franklin. X rays, when a beam of them shines through a crystal, bounce off the layers of identical molecules in the crystal to produce a pattern of spots on the film. Helical molecules, packed in a crystal, produce a unique sort of crossways pattern—and Crick had worked out the mathematics of that pattern and had taught Watson how to recognize it.

The packing of molecules into crystals is for science the most interesting form of the patterns made by packing squares and hexagons and triangles into mosaics. In the nineteenth century, even before chemists were altogether sure what molecules really were, crystallographers understood that any substance that formed a crystal must be made up, at the level of its smallest units far beyond the power of the microscope to resolve, of identical pieces. Crystallographers knew, too, that the visible shape of the crystal—as it revealed itself, for example, when broken along its natural cleavage lines—was trying to say something about the shapes of those smallest units. Most important, a crystal will often be symmetrical in one way or another, sometimes in several ways. It may be identical with itself when given a half-turn, or even a quarter-turn; one part may be the mirror reflection of the other. Snowflakes are the familiar examples, but striking symmetries occur in many crystals more durable than ice. Once more, these reveal the symmetries of the arrangements of atoms and molecules that make up the crystals.

We can see how this might be so even in the patterns of ceramic tiles on a kitchen wall. The great wall mosaics of medieval Christian art—Ravenna or Constantinople—do not display such symmetries because the tiles were chipped and shaped to allow faces and figures to be pictured; for the intricate entertainment of abstract symmetries one must look at the mosaics of the Arabs and Moors. By the turn of our century, the crystallographers had enumerated thirty-two different patterns of symmetry that crystals can show. How to read from those back to the arrangement of atoms and molecules within the crystal took the next fifty-five years to work out fully. But then a Russian crystallographer noticed, to his delight, that the fourteenth-century mosaics of the Alhambra, the high achievement of Moorish architecture in Spain, displayed at one place or another in their patterns all the forms of symmetry of the crystallographers. Forbidden by the Koran from making images, the Moorish artists had been driven to create representations of relationships that underlie all the natural things they saw around them. Once more, restrictions are often essential to the creative leap in art, as in science.

Our recognition of pattern, so quick, so direct, is profoundly hard to explain. Pondering the problem of a theory of visual recognition of pattern, John von Neumann—who had the subtlest and deepest mathematical mind of our century—concluded that the process may be so intricate that we can never adequately analyze it. Actually to build a machine that can recognize patterns as we ourselves do, von Neumann thought, would be simpler and briefer than any adequate written or mathematical description of the process. Diagraming the neural signals received by the cells of the visual cortex of the brains of Siamese cats, David Hubel—a physiologist at Harvard Medical School—found that the elements of perception are built into the interconnections between eye and brain, so that pattern has already been analyzed and in some deep sense recognized even as the neural impulses are routed to and reach particular cells in the brain. Retina and optic nerves and visual cortex actively map the world out there, creating pattern, at every instant—and *how* we map it is partly inborn. But despite years of work, Hubel has been able to drive that discovery no further. And yet agreement about the similarity of patterns is so natural that (in an example I owe to the English physicist John Ziman) people not specially trained in science are employed in particle-physics laboratories to search the thousands of photographs, produced there, of the tracks made by high-energy charged particles in bubble chambers or directly in the film emulsion, to spot the strange configurations, at best extremely rare, that betray the passing or the interactions of new particles predicted by theorists.

Mathematicians have ways to break down and rebuild patterns of every kind. Any complex periodic phenomenon—rhythm in space, pattern in time—that can be measured can be analyzed into a series of simple forms. Music is the model. A pure note, as produced by a tuning fork, will drive a recording pen (as it drives the eardrum) to produce a pure undulating curve, a sine wave. Its frequency represents the tone, its height the loudness. An oboe, a flute, a clarinet, and a saxophone playing that same note are easy to tell apart: though the pure note, called the fundamental, is still there, it has added to it a series of overtones, the harmonics at fractional intervals above the fundamental. These vary in loudness and in timing to produce, all together, the recognizable tone quality of each instrument. But each separate harmonic, in its frequency and loudness, can be represented by a simple sine wave, too. If an electronic synthesizer emits the fundamental tone and every harmonic at the right volume, the result will sound indistinguishably like the original instrument. But the fact is that any pattern of any kind—a picket fence, a mosaic in the Alhambra, the signals received by a radio telescope from deepest space—can be treated the same way, analyzed into a fundamental wave and harmonics. Working the other way, if a series of the harmonics can be measured separately, the pattern that produced them can be reconstructed. These methods, called Fourier analysis after the mathematician of Napoleonic France who first developed them, have proved to be immensely powerful throughout science and engineering. They suggest, once more, how pattern precedes and provokes mathematics.

Often, of course, pattern has led to discovery in less formalized ways. In the 1840s and '50s, cholera emerged from India and swept the world in a devastating pandemic. The disease is peculiarly nasty: the normal passage of liquids through the walls of the intestines is reversed, so that fluid moves from tissues and bloodstream into the intestines and is then lost, by the gallon, in uncontrollable diarrhea. Death follows from acute dehydration. How cholera spreads, or even that diseases may be caused by specific agents like bacteria and viruses, was not at all understood.

But one of the leading physicians of London, John Snow, a pioneer in anesthesia and an obstetrician to Queen Victoria, suspected that cholera was transmitted by water contaminated with sewage. In the summer of 1848, in a section of Soho, in London, around Golden Square and Broad Street, cholera broke out suddenly and more terribly than ever before in England. Within a circle of 500 yards, in a period of ten days, more than 500 people died of cholera. Snow suspected that the cause was a public pump, in Broad Street, whose water was widely used in the area. As the epidemic reached its peak, Snow got the record of the deaths in the first three days of September, which numbered 83. He found where the victims had lived and worked, and investigated where they had drawn their water. He was able to get information for 77 of these cases, and to show that 59 had certainly drunk water from the Broad Street well.

Spot maps of the pattern of infection are now the first tool of epidemiology: the one here, the central part of Snow's map of the deaths from cholera in the neighborhood of the Broad Street pump in the outbreak in the summer of 1848, was the earliest. But some of Snow's most telling evidence came from the apparent deviations from the pattern. A workhouse in Poland Street was almost surrounded by houses where people died of cholera, yet only 5 cases occurred among its 535 inmates. If the mortality had been up to the surrounding area, the workhouse would have had 50 deaths. Snow found that the workhouse had its own well. A brewery stood in Broad Street near the pump, but never used that water, and had no cases among its 70 workers. On Snow's map, only 10 deaths were spotted decidedly nearer to any other pump. "In 5 of these cases the families of the deceased persons told me that they always sent to the pump in Broad Street, as they preferred the water to that of the pump which was nearer," Snow wrote, some years later. "In three other cases, the deceased were children who went to school near the pump in Broad Street." In a conclusive contrast, a "gentlemen from Brighton" came to see his brother, who lived in Poland Street. The brother was already dead of cholera; the visitor did not even view the body, but spent twenty minutes at the house, where he ate a hasty lunch of steak and "a small tumbler of brandy-and-water, the water being from the Broad Street pump." He left the area—and died of cholera forty-eight hours later. A Mrs. E______, who lived in Hampstead, a part of London five miles away, had not visited Soho for many months; but she liked the water from the Broad Street pump, and had a large bottle of it brought to her every day. She drank some on August 31, and died of cholera two days later. "A niece who was on a visit to this lady also drank of the water; she returned to her residence, a high and healthy part of Islington, was attacked with cholera, and died also." There were no other cases of cholera in Hampstead or Islington. What seemed to be violations of the pattern, when Snow examined them closely, turned out to be striking confirmations of it.

On September 7, 1848, Snow persuaded the local Board of Guardians to take the handle off the Broad Street pump. The epidemic was declining, and soon stopped.

In 1853 and 1854, cholera struck London again in epidemic force, this time chiefly south of the Thames. Much of that part of London was then supplied with water by two private water companies—the Southwark & Vauxhall Company and the Lambeth Company—that competed so directly that they often laid their pipes running in parallel in the same streets, so that some houses bought their water from one while neighbors bought water from the other. Snow was able to show, with house-to-house inquiries and a new map, that cholera struck down people who got their water from the Southwark company, while those who drank only the Lambeth water did not get sick. The Southwark water was drawn from a polluted downstream reach of the Thames, while the Lambeth water was drawn far upstream. This is a fine example of the natural experiment; such experiments often must be found in order to advance understanding in areas where directly manipulative experiments are impossible, either for ethical reasons, as here, or because the phenomena are far out of reach, as, for example, in astrophysics.

The organism that causes cholera was not identified until 1883, and how it acts—by producing a toxin that reverses the flow of liquids through the cells of the intestinal wall—was not shown until the late 1960s. But Snow, by his sensational demonstrations of the patterns of transmission of cholera, founded almost single-handed the science of epidemiology. His evidence gave great force to one of the most extraordinary efforts at city planning and public health in history, the Victorian engineering of clean waterworks and effective sewer systems.

The mapping of cholera outbreaks continues to the present day. Cholera is one of those diseases that now must be reported by governments to the World Health Organization. Apart from areas where cholera is endemic, or where epidemics can be traced back along the routes of trade or pilgrimage, WHO records show sporadic clusters or isolated cases with no obvious source—for example, in the central highlands of France. Late in 1978, Dr. Charles Rondle and colleagues, at the London School of Hygiene and Tropical Medicine, offered a bizarre but well-argued hypothesis. On maps of Europe and North Africa spotted with the unexplained cases, they also drew in exactly the usual flight paths of regular air services from the Indian subcontinent to Paris, London, and so on. The pattern of the disease fit perfectly beneath the pattern of these routes. Modern passenger airplanes collect and retain sewage in tanks—except that handbasins usually empty directly into the atmosphere, as may any excess overflowing from the tanks. The bacteria that cause cholera are relatively delicate, and any liquid discharged from a plane at typical long-distance altitudes is frozen in droplets and thawed, perhaps several times, in falling; but Rondle and his colleagues showed by experiments that the organisms could survive such conditions and reach the ground in concentrations that could cause outbreaks. And among air passengers, not only active cases of the disease but carriers without diarrhea can release large numbers of the bacteria by washing their hands. Rondle and his co-workers ended with a nasty warning: "If cholera can be spread, even only occasionally, by effluent from aircraft, then close investigation should be made of the possibility of other bacteria and viruses being spread the same way."

The most tremendous perception of pattern in the history of science, though, was Dmitri Mendeleev's creation of the periodic table of the elements. Mendeleev perceived the pattern first in 1869, in the midst of data that were conflicting, incomplete, and often erroneous. The pattern brought order into chemistry at the most fundamental level. The pattern has grown steadily in explanatory power—often in ways that Mendeleev himself did not anticipate, or even disagreed with. Mendeleev not only perceived the pattern, he *imposed* it—insisting, in case after case, that when the known facts of chemistry did not fit his pattern, the facts were at fault.

That substances are made up of atoms had been established by 1803 by John Dalton, in England. Through the middle third of the nineteenth century, chemists were sorting out the ways and proportions in which different substances combine with one another. They were determining which of them were the elements out of which other substances were built, measuring the elements' atomic weights relative to hydrogen, the lightest, and characterizing their chemical behavior. By the mid-1860s, about sixty different elements had been identified, from hydrogen through uranium. Yet chemistry was a chaos.

Thoughtful chemists were tantalized by glimpses of order in the chaos. It was obvious, to begin with, that some elements widely different in atomic weight nevertheless ought to be related. For instance, sulfur, selenium, and tellurium, although their atomic weights were 32 (hydrogen equaling 1), 79, and 127.6, all were nonmetallic, formed bright-colored crystals, and combined with hydrogen in similar proportions to make very smelly substances that were weakly acid. Several people proposed tentative groupings of elements based on related properties. In 1864, the English chemist John Newlands tried arranging the lightest of the elements then known in order of increasing atomic weight. Newlands saw that the eighth element in his table, fluorine, had properties somewhat like the first, hydrogen, as well as the fifteenth, chlorine; that the twelfth, silicon, had properties like the fifth, carbon—and so on, in a crude way. The repetition of properties eight places apart in the series tempted Newlands to compare these chemical periods with musical octaves—and to assert that he had found a fundamental harmony in nature. He was heard with indifference, even ridicule. His groupings indeed had grave flaws. He left no way to place elements as yet undiscovered. He had not scrutinized critically the work done by others on atomic weights, to pick the most reliable value for each element. Worst, certain of his octaves were not harmonious: for instance, sulfur, not a metal, hardly resembled iron, nor did phosphorus, not a metal, resemble manganese.

Boldness is an aspect—often the crucial one—of scientific genius. Mendeleev was born in Siberia, the youngest of seventeen children; at the age of thirty-three he was appointed to the chair of chemistry at the University of St. Petersburg. He found that no good textbook existed for his students, so he began to write one—*Principles of Chemistry*. His own research had taken him deeply and critically into the chemical properties of the elements and their correlation with atomic weights. By then, too, atomic weights for the elements long known had been revised or refined, and fairly accurate atomic weights had been proposed for many of the new ones.

On March 1, 1869, Mendeleev was about to leave St. Petersburg on a trip to the provinces—to look into methods of making cheese—when he realized suddenly that, if he arranged the elements according to increasing atomic weight, similar chemical properties reappeared cyclically. He had not then heard of Newlands's octaves. Leaving hydrogen aside, as unique, Mendeleev first essayed:

	Fluorine (atomic wt. 19)	Chlorine (35.5)	Bromine (80)	Tellurium (128)
Lithium (7)	Sodium (23)	Potassium (39)	Rubidium (85.4)	Cesium (133)
		Calcium (40)	Strontium (87.6)	Barium (137)

The second line of that table lists alkali metals; the elements in Mendeleev's third line are now called alkali earth metals. Their nearest kin in chemical behavior, Mendeleev then saw, were the transitional metals copper, silver, and gold. These, Mendeleev realized that day, could be placed in the table as a new line—

Copper (63.4)	Silver (108)	Gold (197)

—with the weights falling in the right ranges. He quickly added other elements singly and in groups.

As quickly, he began to use the developing table in a way that precursors like Newlands hadn't dreamed of, to criticize and correct the chemical information he started with. In that month of March 1869 Mendeleev announced the periodic law: "The elements, if arranged according to their atomic weights, exhibit an evident *periodicity* of properties." His earliest papers had errors: he put lead at the end of the group with calcium and barium, for instance, and uranium with boron and aluminum.

Mendeleev liked playing solitaire; now, as he worked out the relations of the elements, he put each one on a blank card, with its atomic weight and chemical character, then arranged and rearranged the cards. The table evolved rapidly toward the form used today. By 1871, when it was first published in English, sixty-three elements were listed—including many, like platinum, chromium, and others, where Mendeleev called for new determinations of atomic weights so that they could fit where he thought they belonged. Uranium, for instance, he said could not have the atomic weight of 120 that was generally accepted, but must be 240.

The table had blanks—for, in the boldest assertion of all, Mendeleev left holes where no element was known that matched the weight and properties he thought belonged there. Thus, to fill a space below silicon and above tin, he called for a new element and named it *ekasilicon.* Its atomic weight should be near the average of silicon and tin. Its chemical properties he predicted from the comparison of silicon with its immediate neighbor, phosphorus, and of tin with its neighbor, antimony; ekasilicon should show a similar relation to *its* neighbor, arsenic. Similarly, he left blanks for *ekaboron* and *eka-aluminum.*

Opposition to Mendeleev's ideas at first was vigorous and widespread. From 1871 to 1874, however, experimental work in laboratories in many parts of Europe produced corrected values for atomic weights of various elements that again and again agreed with what Mendeleev had predicted.

Then in 1875, the French chemist Lecoq de Boisbaudran discovered a new metal and named it *gallium.* Boisbaudran worked in ignorance of Mendeleev's table. Gallium, though, had properties like those Mendeleev predicted for eka-aluminum—except that it was too light. Mendeleev sent off a brief paper to claim that gallium was eka-aluminum. Boisbaudran objected. Then he measured gallium's atomic weight again, and found that he had originally figured it incorrectly. Mendeleev's predicted weight was right. His pattern had forecast the properties of the new element more accurately than they had been determined by the man who actually discovered it. The world of chemistry was astounded. Six years after Mendeleev first saw his periodic table, it was firmly established. In 1879, Lars Fredrik Nilson, in Sweden, found a new element he named *scandium:* Mendeleev's ekaboron. In 1886, Clemens Alexander Winkler, in Germany, announced *germanium:* its properties matched ekasilicon with amazing accuracy.

Twenty years after Mendeleev's announcement of the periodic law, he was invited to give the Faraday Lecture before the Chemical Society, in London. For that lecture he wrote, in part: "The law of periodicity first enabled us to perceive undiscovered elements at a distance which formerly was inaccessible to chemical vision. . . . When, in 1871, I described to the Russian Chemical Society the properties, clearly defined by the periodic law, which such elements ought to possess, I never hoped that I should live to mention their discovery to the Chemical Society of Great Britain." Then he predicted that still more elements would be found—and described the properties of one he called *dvi-tellurium.* In 1898, the element that matched the prediction was isolated by Marie and Pierre Curie, in Paris, and named *polonium* after her native Poland. Polonium had one property Mendeleev in 1889 could not have predicted: it was radioactive.

A Conversation with George Alcock

The greatest master of the patterns of the sky—an amateur.

Since men and women first looked up at the sky at night, the stars have entranced us. The patterns we mark out as the constellations and teach our children on a crisp fall night—Look! Orion's Belt!—have been passed down to us from the Romans, from the Greeks, from the Sumerians and before. The Mayans, the Chinese, marked the constellations differently, but mark them they did.

Always we have watched the night sky for interlopers. Against the wheeling backdrop, the planets—the wanderers—traced long cycles that could with years of observation be worked out. Meteors were as unpredictable as the weather (and thought to be related, which is why their name shares a Greek root with the science of weather). Yet for these, too, the records revealed patterns. Showers of meteors appear at the same times every year, like the Geminids in mid-December, the Perseids—most spectacular of meteor showers—at the twelfth of August. Comets were rarer, slower—and, with their size and streaming hair, very weird. They foretold the death of kings. Shakespeare places a comet in the sky before the murder of Julius Caesar, and the Bayeux Tapestry, telling the story of the Norman Conquest like the world's first and largest comic strip, puts in one frame the comet that appeared in 1064. Not until Edmund Halley, Newton's publisher and himself an astronomer, observed the great comet of 1682, did anybody notice the pattern of dates and realize that some comets return. The Bayeux Comet was Halley's. These are members of the solar system, though on a very long leash. And then, defying all prediction, there are the new true stars—the novae and the rare supernovae—that flare up suddenly and die away.

The all-time master of the patterns of the stars is an amateur: George Eric Deacon Alcock, a schoolteacher in England. Alcock tracked meteors by eye in the 1930s and was one of a team of amateur observers who were right about their paths and origins when the most eminent professional astronomers were wrong. In 1951 he started to search the sky with binoculars for new comets, and has now spotted four—a record, for only about six hundred different comets are known ever to have been seen. Then he began to memorize the patterns of the stars, through binoculars, so completely and perfectly that he could catch new stars flaring up. When he started looking, although some three hundred novae had been recorded since history began, only about nineteen had been observed satisfactorily.

Alcock is of medium height, stocky, in his late sixties, gruff and matter-of-fact. He lets one know that he's had his battles for recognition. He lives with his wife in a small house just outside Peterborough, a city of England's industrial Midlands; he taught in Peterborough for more than forty-eight years until he retired in 1977. Every night when the visibility is good enough, he takes a folding chair into his tiny back garden, leans back with his elbows supported for steadiness, and sweeps the sky. The dimmest stars visible with the naked eye in perfect seeing conditions—on some cloudless mountaintop far from dust and city lights—are those of the sixth magnitude. Perhaps six thousand stars of the sixth magnitude or brighter can be seen on a perfect night in the northern hemisphere. Alcock's binoculars allow him to see many times that, to magnitude 8½.

I asked him how and why he did it.

"I became interested in astronomy almost by accident," Alcock said. "There was a little encyclopedia, distributed by some soap manufacturer, and in the front it had a chart—a 'planisphere.' I was about ten years old. When I was eleven or twelve, Rutherford split the atom, and frightened everybody. And there was a close approach of

Mars in the news. And then a small solar telescope came my way; it had a good dark filter—no danger to the eyes—and I was able to look at the sunspots in the 1926 maximum. I suppose that capped it.

"Our meteor work turns out to have been important for the astronauts. Some friends and I—we were all amateurs—began in the early thirties. I had learned off the night sky by eye, even before we began. We found out the heights at which meteors *start* to burn out, and the heights at which they *finish* burning out. In that work, we were discovering things that were never discovered before. There was a handful of us. You learned the night sky by the numbers in the star charts, down to the fifth magnitude. All the sky, as a backcloth. Because you've not got time to look at the charts—you'd spend more time inside the house than out!" He and friends in towns tens of miles away, watches synchronized, observed meteor tracks, recording start and end points by the numbers, then compared their notes to triangulate altitudes, speeds, trajectories. "It was *great fun.* And we said that by their speeds, meteors almost always were on elliptical or parabolic paths. That meant they were part of the solar system. But then at the end of the thirties the great Harlow Shapley and others began to turn professional instruments on meteors, and *they* came to the conclusion that most meteors were on hyperbolic paths and so were coming from *outside* the solar system. And I couldn't fight a name like Shapley if I was a backyard astronomer. But after the war, when at last radar began to be used to track meteor paths, they were indeed found to be in orbits in the solar system. So the amateurs, with our naked-eye methods, were right after all."

In 1951, Alcock began searching for new comets. "Three years later, I stopped to think that after all there were about one hundred and seventy nights a year here that were clear enough to do anything—so I'd make better use of the time if I had another string to my bow. I didn't know much about novae, but a friend of mine had found Nova Herculis 1934—which was the first nova of my lifetime. Between 1934 and 1950, there had been only three naked-eye novae. I began to read about them, and learned that they were roughly divided into two kinds, fast and slow—that is, the ones that died away quickly and the ones that lasted awhile.

"That was what set me off to see if I could learn off any part of the night sky, thoroughly enough—down to magnitude 8½. In the band of the Milky Way. You don't look at a star! You look at patches of stars—patterns. This patch looks like a kind of bird, that one like a plane. I started learning at the constellation Cygnus and moved down the Milky Way to the southwest—to Sagittarius. Then the other way to Canis Major. The sun covers a lot of that area—that is, the stars go behind the sun for part of the year. I was afraid I would have to learn them all over again when they came back out! It took me about six years before I felt I knew enough even to *begin* the patrol for novae."

While he was memorizing the sky he found his first comet, on August 24, 1959—and six nights later, his second. He also found hundreds of errors and omissions in the official star charts and catalogues. In 1961 he began scanning for novae in earnest. He got his third comet in 1963, his fourth in 1965. And he was still extending his knowledge of the patterns of the sky, out to the sides of the Milky Way. "I now cover the Milky Way band at a width of thirty degrees each side of the galactic equator—a swath of sixty degrees all told. How many stars? I can't say. But if I estimate I look at four hundred a minute—it takes me an hour and a half to sweep the sky completely." Alcock looks at nearly forty thousand stars on a clear night. "Then if there's a *stranger* in their midst, at any point, something different, something that bothers me—I check the charts. Perhaps I get up to check three or four times a night. There are quite a lot of variable stars to give trouble. But I've learned most of those.

"It wasn't until 1967 that I found the first. July eighth,

1967. In a small constellation called the Dolphin. *Instantly* I knew I had it. Instantly knew there was a stranger. You sit for a second wondering. And then you get on the phone as quickly as you can. To the director of the comet section of the British Astronomical Association." Nova Delphini 1967 was at magnitude 5.6 when Alcock first spotted it. A coded cable went from the Astronomical Association to the Smithsonian Institution in Washington, which operates a clearinghouse and notifies the world's observatories—the Japanese, first, because the dark is moving that way, then the Europeans. Nova Delphini was a very unusual new star. It rose slowly and irregularly to magnitude 3.5 in December, easily visible to the unaided eye, and took more than a year to fade away, far longer than typical. "The first one, they said I was just lucky." But Alcock found his second nova on April 15, 1968, in the nearby constellation of Vulpecula. "The second one showed up only ten degrees away from the first one. So they said, 'Well, that's the only part of the sky he knows.' That was a fast nova—but for several weeks there were two naked-eye novae visible at the same time. Unprecedented. And both were found by an Englishman! Like Edmund Halley, I am very patriotic." Alcock found his third nova—a record for visual observation—at the end of July of 1970. "It was the faintest ever discovered visually—though fainter ones have been found on photographs—and was well away from my others. At that point I think they began to believe me!" One night in 1971, Alcock spent over an hour scrutinizing a patch in the constellation Cepheus, "Because there was something the matter with it." But he couldn't pinpoint the discomfort. A day or two later, a Japanese astronomer announced that a nova there had been discovered on a photographic plate. Alcock waited six years for his fourth nova, again in the constellation Vulpecula, on October 21, 1976.

I asked Alcock what the experience was like when he first began to memorize the patterns of the stars—it must have been extraordinarily daunting, extraordinarily beautiful. "Oh!" he said. "It still is! Looking through binoculars you see *more* of the sky. Up to a thousand stars at once, on a clear night, in one field of view. It's an *amazing* sight."

3

Change

Shakespeare, following the propaganda of the Tudor historians, painted Richard III as a humpbacked Machiavellian monster. It's unquestionable that Richard was a small man, as well as a warrior of extreme skill, bravery, and ferocity—the greatest fighter and commander of his time in England. Did Richard have an advantage in armored combat because he was short?

That suggestion was made some years ago by one of the leading modern historians of the Tudor era, Garrett Mattingly. In a lecture on the beginnings of the Tudor dynasty (and propaganda) at the battle of Bosworth, Mattingly pointed out that between a short man and a tall man, height increases by the linear dimension—from 5 feet 2 inches, say, to 6 feet—while the surface of the body increases as the square. Since it's not just the length but the surface of the body that the armorer must plate with steel, the armor of a short warrior, like Richard, would be lighter than a tall warrior's by a lot more than the few inches' difference in height would indicate. So Richard's notorious deadliness in battle would have been possible at least in part—or so Mattingly's speculation ran—because his armor, while protecting him as well as the big man's, left him less encumbered.

The principle Mattingly appealed to goes back to Galileo and explains an astonishing variety of puzzles in science and technology: why flies can walk on the ceiling, why the tallest tree cannot be more than 300 feet high, why a child can take a tumble that would break bones in an adult, why shrews are cannibals, why a large girder is proportionally weaker than a small one. Length is linear, surface is square, and volume—as Mattingly neglected to add—is cubic. So when you double the linear dimensions of a simple solid object, its surface increases not twice but the square of twice, or four times, and its volume goes up by the cube, or eight times. The principle is called the scale effect. But does it really help explain Richard III's success in battle? After the lecture, someone said to Mattingly that he had grasped the right idea—but by the wrong end.

Muscle power, the listener claimed, is a matter of bulk—and physical volume goes up by the cube, while the surface to be protected goes up by the square. So the large warrior should have more strength left over after putting on his armor than does the little guy. And the large warrior, swinging a bigger club, can deliver a far more punishing blow—because the momentum of the club depends on its weight, which goes up, once again, with its volume, which means by the cube. Richard was at a terrible disadvantage.

But wait a minute, a second listener said. That's true about the club—but not about the muscles. The strength of a muscle is proportional not to its bulk but to the area of its cross section. And since the cross section of muscles obviously increases by the square, just as does the surface of the body, the big guy, plated out, has no more, or less, advantage over the little guy than if both were mother naked.

But hang on, a third person interjected—an engineer. That's right about the muscles, but it's not right about the armor. The weight of the armor increases not simply with the increase in surface area that it must cover, but slightly faster. The reason is that to obtain the same strength with a larger area of metal, there must be reinforcing ribs. Or else the metal must be significantly thicker, overall. So maybe Richard had an advantage after all.

At which moment, Mattingly reminded us that Richard III was the last king of England to die in battle—and that he did so in the attempt personally to cut down his rival, Henry Tudor. Which introduced what are, after all, the most important variables of them all: skill and strength of will.

The problem of the armorer was to maximize protection yet maximize mobility as well—at a weight that could be borne. In antiquity, the top load for an able-bodied slave in a pack train was about 56 pounds; overseers knew that given a heavier burden the slave would not be able to get up and go on the next day. In the Middle Ages, a knight's full suit of armor rarely weighed more than about 56 pounds. Armor evolved over centuries. The armorer could never escape the fundamental relation of the three-dimensional body to its two-dimensional surface that was to be protected. He could never evade that same relation when it showed up as the ratio of surface to thickness in the strength of the materials he used. But the armorer's responses were subtle and artful. A knight's armor was tailor-made for him. At the height of the fifteenth century, a full fighting suit comprised more than two hundred pieces. A gauntlet alone was put together from more than a score of pieces, articulated to slide over each other in movement. Put on a medieval gauntlet and you will experience an extraordinary sensation of lightness and flexibility—as though, for a moment, your hand is transformed into a snake. Armorers were such knowing anatomists that modern makers of artificial legs have learned from ancient armor how to construct a knee joint that adjusts correctly with the changing relation, as the leg swings, of the length of the thigh to the length of the lower leg, to give a smooth gait.

To maximize protection within the weight, the armorer adopted two strategies: variable thickness and deflection. The unexpected fact is that armor was made as thin as possible. Thickness, reinforcement, structural stiffening of a large surface, were concentrated where opponents' weapons were likely to hit—across brow and temples and down the middle of the face, across shoulders, at elbows, down the center of the breast, at knees and lower legs where a mounted man was most exposed to the shock of collision with another or to the slash and stab of men on foot. At these critical points the metal was shaped not merely to resist but to deflect the blow so that the point or cutting edge would slip away and off. From these strong, shaped places, the metal tapered away—until at the sides of the rib cage beneath the arms, or across the fingers, or at the cheek of a helmet, the sheet steel was as thin as the modern lid of a coffee can.

The scale effect, and particularly the surface-to-volume ratio, pervades science. A great part of engineering is devoted to evading the consequences of the scale effect or putting them to use. No other set of relationships is as simple yet as fundamental. The changing relation of surface to volume creates domains so different, especially among living creatures, that the laws of nature seem almost to be altered in their application.

The first difference that defines these domains is that gravity acts on a creature's mass, which goes up with its volume, while many other effects—including those of wind and water—act on its surface. For this reason, some familiar beasts experience the world very differently from the way we do. Years ago, in an amusing essay, "On Being the Right Size," the British biologist J. B. S. Haldane wrote: "You can drop a mouse down a thousand-yard mine shaft; and, on arriving at the bottom, it gets a slight shock and walks away. A rat would probably be killed, though it can fall safely from the eleventh story of a building; a man is killed, a horse splashes." The difference is air resistance—which is proportional to the surface of the moving object. Haldane went on, "Divide an animal's length, breadth, and height each by ten; its weight is reduced to a thousandth, but its surface only to a hundredth. So the resistance to falling in the case of a small animal is relatively ten times greater than the driving force."

In that domain, the size of the mouse and smaller, gravity presents few dangers. The housefly walks on the ceiling not because the pads of its feet are tiny suction cups, nor because they exude some sticky stuff, but by a much simpler means—and yet the means is all but beyond the direct experience of creatures our size. Nowadays, engineers make critically exact measurements with electronic devices, but in the years before such tools were invented, the most precise sort of gauge for certain engineering measurements consisted of a set of steel blocks, called Johansson blocks. These were of assorted sizes up to about 6 inches long, surfaces so flat and so smoothly polished that when two of them were wrung together to make up a specific length, they could not be pulled apart again. The supersmoothness allowed the molecules of the two surfaces to get so close that intermolecular attraction bound the blocks. The only way to separate them was to slide them apart. But the housefly is so light that the bottoms of its feet, being reasonably flat and flexible, provide enough intermolecular attraction where they touch the wall or ceiling to hold the insect in place. Again, a water beetle is so light in comparison to the surface of its feet that it can step on the water without indenting it enough to overcome the surface tension between the water and the air, and break through. The water bug's ability, unlike the fly's, is in reach of our understanding through analogous personal experience: we can use snowshoes. If the bug were much bigger, or if it tried to skate on a pond of alcohol or gasoline—liquids that have less surface tension—it would penetrate the surface and drown. For very small creatures, body surface is so great in proportion to volume that surface effects present dangers to them as gravity does to us. D'Arcy Thompson, the master of pattern in living things, was first to point out that "a man coming wet from his bath carries a few ounces of water, and is perhaps 1 per cent heavier than before; but a wet fly weighs twice as much as a dry one and becomes a helpless thing. A small insect finds itself imprisoned in a drop of water, and a fly with two feet in one drop finds it hard to extricate them."

In a few species of insects, adult individuals

grow slightly larger than the smallest mouse. That's the upper limit of the insects' domain—for many reasons, most of them variations on the one theme of surface to volume. Insects absorb oxygen not by means of lungs or gills and transport by the blood; instead, they have countless minute tubes that open at many places on the surface of their bodies and permit air to reach the tissues directly. Make the largest insect much larger still, and its respiratory system would fail—for the volume of tissues needing oxygen, and the distance from the surface to the inmost tissues, would require elaborate branching networks. Diffusion of oxygen through the tubes would be slow. They would be pinched with every movement. Many insects, also, wear their skeletons on the outside. As they grow, periodically they shed their external skeletons. Their tissues, unsupported while a new external skeleton forms, if too great in volume would collapse in a blob. The shell itself, if it were much larger, would require thickening or ribbing or ridges—for the lobster's armor can be described in the same terms as the medieval knight's. Further, the beautifully delicate legs characteristic of insects are impossible at proportionally larger sizes—for the strength of a supporting limb, its ability to resist being crushed by the load, increases only with its cross section, while the creature's weight, as always, goes up with its volume.

If an ant blown up to the size of a horse makes a silly horror movie and preposterous engineering, a warm-blooded creature shrunk to the size of a fly is just as impossible. In our own domain, the lower boundary of size is set by the need to keep warm. Heat is generated in the tissues, but it is lost through the skin. Mammals and birds require a food supply related to their surface rather than to their weight. A grown man needs about 1 per cent of his body weight in food each day. A mouse eats a quarter to a half of its body weight in food each day. A hummingbird consumes more than its own weight each day—and must live on nectar, of all natural foods the most rich and concentrated. A shrew, the tiniest mammal—an adult pygmy shrew is smaller than 2 inches, except for its tail, and weighs less than a fifth of an ounce—must eat almost without stopping. In order to survive at that lower boundary, where the surface-to-volume ratio is so disadvantageous, shrews are startlingly vicious. They will kill and eat mice. They will kill and eat each other. In an experiment, three shrews were placed beneath a glass dome with no food. Two promptly attacked and ate the third. Eight hours later, only one remained.

These relations continue up the scale: a 4½-pound rabbit burns about 22 calories per pound, in twenty-four hours; a 150-pound man burns about 15 calories a pound in a day; an elephant, at 4 tons, burns 5 or 6 calories per pound—and a whale, at 150 tons, probably burns less than 1 calorie a day for every pound of its weight. The fat person trying to lose weight may well think the surface-to-volume ratio unfair: just to stay even, he needs proportionally less food than the skinny person—a small but significant difference which is heightened by the fact that a layer of fat beneath the skin is an efficient insulator.

The wings of hummingbirds beat 50 to 75 times a second. Although the birds can hover, when they dart they may attain 60 miles an hour—for a few yards. Bees' wings beat 190 times a second. The heart of the masked shrew beats 800 times a minute. For creatures that size, time itself runs differently. A hundredth of a second makes a difference; a second is filled with events; generations and lifetimes pass in weeks or months.

D'Arcy Thompson, stepping wet from his bath, reaches for the towel rail. Just as the varying thickness of the metal of a cuirass is an engineered response to the problem of the surface-to-volume ratio, so is the construction of the Turkish towel with which D'Arcy is now rubbing down. The cotton fibers of the towel are not individually more absorbent than any other cotton—but the thousands of loops multiply by thousands of times the area of surface available to water. Under a microscope, a cross section of the intestine of any sizable animal reveals a pattern strikingly like a slice through a Turkish towel. Similar structures for similar problems. A worm, say, can have a gut that is a smooth, straight tube. But an animal ten times as long, ten times as tall, ten times as wide—a cat, say—has a thousand times the bulk to be nourished; and yet the interior surface of the worm's smooth, straight tube, similarly enlarged, would have grown only a hundredfold. The cat would starve no matter how fast it ate. The surface of intestines, to be able to pass enough nutriment through, must increase as rapidly as the bulk of the animal. In fact, of course, intestines grow long, so that they must be folded up inside a large creature, and they develop thousands of tiny protuberances—a thick plush that has exactly the same function as the pile in Turkish toweling, to increase by hundreds of times the amount of surface available to absorb liquid. The development of gills or of lungs serves an exactly comparable function—for the surface available to absorb oxygen must grow convoluted to keep pace with the growing bulk of the animal's tissues. A man's lungs have 100 square yards of surface: that's half a tennis court.

Plants, too, respond with a similar strategy to a similar necessity. A daffodil bulb puts up a dozen flat green spears and gets the sun's energy it needs; a silver maple 47 feet tall, 5 feet around the base, requires 175,000 leaves, with over a sixth of an acre of surface, to thrive as well as the daffodil. "A leafy wood, a grassy sward, a piece of sponge, a reef of coral, are all instances of a like phenomenon," D'Arcy Thompson wrote. Our brains are big, though elephants have bigger; but whereas the brains of most animals are relatively smooth, the human brain has always fascinated neurobiologists by the deep and intricate folding of its surface. The conclusion has long seemed inescapable that the human brain requires size and at the same time a high ratio of surface to volume—but the exact significance of this clue for our intelligence and behavior remains baffling.

News reports, every so often, tell of a whale that has driven itself ashore and lies panting on the beach—where, unless promptly towed out to sea again, it dies. The first person to demonstrate the scale effect, and, in passing, to explain how the whale can reach ten times the size of the largest land animal, was Galileo Galilei. In his seventies, ill and going blind, living in seclusion enforced by the Inquisition because of his defense of Copernicus's view that the earth is a planet like the others, moving around the sun, Galileo wrote his last book, *Dialogues Concerning Two New Sciences.* Published in 1638, it set forth most of his work in physics. Galileo was as brilliant a popularizer as he was a scientist—which, indeed, was why the Inquisition feared him. He cast his physics in the form of a four-day discussion among three characters: the sage and didactic Salviati, the attentive Sagredo, and Simplicio—the one who catches on more slowly than the reader. On the second day, Salviati talks of beams and bones. He explains the scale effect with geometric diagrams of the shifting relations of weight to length, and so to strength, in a cylinder—a column or a beam. He proves the crucial point that resistance to crushing or bending varies with the cross section of the cylinder, while the weight to be resisted goes up, of course, with the volume. With a famous drawing, Salviati shows two bones—the second one just three times longer than the first, but then made thick enough to perform as well for the large animal as the smaller bone for the smaller animal. "You see how disproportionate the shape becomes," Salviati says. In a famous passage, he explains:

> From the things demonstrated thus far, there clearly follows the impossibility (not only for art [that is, engineering] but for nature herself) of increasing machines to immense size. Thus it is impossible to build enormous ships, palaces, or temples, for which oars, masts, beamwork, iron chains, and in sum all parts shall hold together; nor could nature make trees of immeasurable size, because their branches would eventually fail of their own weight; and likewise it would be impossible to fashion skeletons for men, horses, or other animals which could exist or carry out their functions proportionably when such animals were increased to immense height—unless the bones were made of much harder and more resistant material than the usual, or were deformed by disproportionate thickening, so that the shape and appearance of the animal would become monstrously gross.*

The delicacy of the leaping fawn is translated by the iron law of scale into the lumbering thickness of the rhinoceros. The elephant, at the extreme upward boundary of our domain, has leg bones partially hollow—built the way an architect designs supporting columns in a massive structure in the attempt to get strength at the least cost in weight. No tree can be taller than three hundred feet, Galileo estimated (and more modern calculations agree). Taller than that, and at the slightest swaying away from the perfect vertical, the tree's own weight would crush it to the earth as the wheat stalk bends under too heavy an ear of grain. And the giants of myth and poetry are impossible. At each step, their thigh bones would break. "On the other hand it follows that when bodies are diminished, their strengths do not diminish in like ratio," Salviati says. "A little dog might carry on his back two or three dogs of the same size, whereas I doubt if a horse could carry even one horse of his own size."

---

*Galileo Galilei, *Two New Sciences*, tr. by Stillman Drake (Madison: University of Wisconsin Press, 1974), pp. 127–28.

But then, Simplicio asks, what about the whale—ten times the size of the elephant?

To which Salviati replies—and the observation was simple, new, and revolutionary—that great size can be reached in only two ways. If the structural materials can be made stronger, size limitations, though never eliminated, are pushed higher. Or if the materials—those supporting and those supported—can be made lighter, the limits are pushed higher. And fish live in a way that makes them effectively much lighter, that eliminates the penalties of weight.

Simplicio suddenly comprehends: You mean because they swim in the sea?

Yes, Salviati says: What happens with marine animals is the reverse of what we know on land—for in the sea it is the buoyancy of the creature's flesh that supports not only its own weight but that of the bones. Inevitably a beached whale quickly dies.

These simple relationships of length to surface to volume-and-weight: since Galileo first opened their endless ramifications to the understanding of Simplicio, D'Arcy Thompson, Haldane, and the rest of the world, scientists have pursued them with fascination. The shifts in these relationships appear everywhere, of course—but more than that, they illustrate the most fundamental necessity in scientific reasoning. Length, area, volume—changeable elements of a problem, like these, the scientist calls parameters. Temperature, pressure, tensile strength, concentration, many other elements in a given problem may be its parameters. I asked Murray Gell-Mann how he defines *parameter*. His answer: "a variable constant." The definition is itself thought-teasing: the parameters of a problem are those elements by which all the rest can be expressed—the things that, when changed, make a difference.

Thus, whenever a scientist tries a series of

experiments, he must first identify what the parameters of the problem might be. Only then can he set up the situation so that he can change each parameter, one at a time, while holding all the others constant. To expand Gell-Mann's definition, a parameter is something the scientist must hold constant except when he is deliberately varying it.

Galileo's most extraordinary contribution to the founding of modern science was to show others, not only with the scale effect but in scores of demonstrations and experiments, how to think clearly about parameters. His diagrams and examples of the scale effect have another charm: they are early, pure examples of what in this century we have learned to call a *Gedankenexperiment*—a thought experiment. Galileo was far from the static spirit of the theorems of Euclid: Galileo's diagrams are dynamic, physical, asserting that "if you change this, in this direction, here is what then happens." The most famous thought experiment in history is the one in Albert Einstein's paper in 1905 that introduced the theory of relativity: he began a revolution with the mild admonition, "When I say, for example, 'The train arrives at Zurich station at seven o'clock,' what that means is, 'The arrival of the little hand of my watch at the number seven and the arrival of the train are simultaneous events.'"

A Conversation with Helmuth Nickel

Were the knights in armor, our medieval ancestors, smaller men than nowadays?

If one looks at the suits of armor on display in one of the great collections—at the Metropolitan Museum of Art, in New York, in the Tower of London, in the Armor Collection in Vienna—one sees that many of the suits would be painfully hard to get into. Indeed, the belief is widespread that armor shows that the fighting men of the Middle Ages were smaller than men today. In the course of a recent conversation with Helmuth Nickel, curator of arms and armor at the Metropolitan Museum, we asked about that. His reply was almost indignant.

"That is one of those stories *everybody* tells me—and it simply is true and yet *not* true," Nickel said. "For example, if you go to the Costume Institute—or any collection of historical costumes—you will find there plenty of tiny shoes, tiny dresses, tiny everything, suggesting also that people used to be tiny. But why is that?

"If you have expensive clothes, and you are a person of normal size, you wear them; and if they are not worn out, if they become unfashionable you pass them on to someone else, a younger brother or someone. That way clothes of normal size get totally worn out. And things of *larger* size, unless they are Peter the Great's huge boots or something, get cut down to fit when they're passed on, whenever there's a way to do it. But things of small size you cannot do anything with—and so they are the things that survive and are now in costume collections.

"That's why I believe the warriors of the Middle Ages were of normal size compared to now. If you have a suit of armor, it's a major investment. The first suit of armor you got when you were freshly knighted, when you were seventeen or so. Then by the time you are twenty-two you've outgrown that. But it's too good to throw away, so it hangs in your armory. You get a new larger suit—and *that* one you wear for some years, and if it's still wearable when you next get a new one, it's passed on to somebody else and gets worn until it's full of holes and one by one the pieces are thrown away. And if you are killed, somebody else also will wear your armor. But meanwhile, that first suit from when you were seventeen is still hanging in the armory. And if you pass that on to your son when *he's* seventeen, he'll say, 'That's hopelessly out of date—I can't be seen by my friends in *that* old thing.' So nobody uses it again. And when you are old and retired, ninety years old and sitting there by the fireplace, your grandchildren will always recall that their grandfather was a little old man—and there's his small armor hanging in the armory to prove that he was such a little man. In the meantime, in your prime, you were five feet ten—but nobody remembers.

"And, by the way, it's not true either that suits of armor were so heavy that if a knight was knocked off his horse he couldn't get up again. That would have left him at the mercy of the foot soldiers. A fully armored knight carried into battle about the same weight that a soldier in the Second World War carried."

4

Chance

Flies swarmed around the urine of a dog whose pancreas had been removed, and a laboratory assistant pointed them out. Sets of fresh frogs' legs, strung on a copper wire and hanging near an iron balustrade, twitched convulsively when the wind brushed them against the metal, and a physiologist thought that peculiar. A rabbit's ear was too warm, when an experimentalist had expected it to be cold. A photographic plate, wrapped to keep out light and put away with a pinch of fluorescent powder for several days, on developing was found to be deeply darkened, when the slightest darkening or none at all was anticipated. Some chickens were inoculated with a culture of bacteria that happened to be stale, and thereupon sickened yet recovered when they weren't supposed to. An animal that had survived a large dose of the toxin of the sea anemone was given a second but far smaller dose a fortnight later, and died. A colony of green mold, contaminating a petri dish where staphylococci were growing, was surrounded by a circle of dead, burst bacteria.

How chance is chance? Of the seven encounters with evidence just itemized, every one was a happenstance, a surprise. Every one led to a discovery of highest importance. Indeed, three of the discoveries led to Nobel Prizes while the other four were simply too early for that. Yet there's a genuine conundrum here, both logical and psychological. New science must be surprising: if an idea follows from what's already known, though it may be a valuable corollary it is redundant. "It is impossible to predict new ideas—the ideas people are going to have in ten years' or in ten minutes' time—and we are caught in a logical paradox the moment we try to do so," Sir Peter Medawar once pointed out. "For to predict an idea is to have an idea, and if we have an idea it can no longer be the subject of a prediction." Beyond even that, though, what a scientific model or theory does that is most profoundly persuasive, the act by which it turns from a clever idea into a program of research, is to predict the discovery of surprising, previously unexpected new facts. Thus, at every turn, science is intrinsically surprising. It is, so to speak, predictable that some discoveries will be made by chance. Yet when it happens, when the surprising observation precedes the model or the theory so that the discovery seems to be stumbled upon, the event is exciting, of course—and unsettling. It raises questions: What have we missed? How do we recognize a significant fact unless to some degree it fits into a pattern of other evidence and a tissue of ideas? Was the discovery therefore inevitable, the times right for it, with the chance residing merely in the particular circumstances?

Tales of chance discovery cut deftly and painlessly to the deep structure of the scientific process; not only laymen but scientists find them irresistible. "In the field of observation, chance favors only the prepared mind," Louis Pasteur once wrote—a famously haughty remark, forever quoted and misquoted, but one of reverberating emptiness. It doesn't say how. A surprise observation that falls upon a prepared mind can be compared, perhaps, to the punch line of a joke: what was expected to be the conclusion of one pattern is suddenly seen to be part of a different pattern that was building, all but unnoticed, alongside the first. Lewis Thomas, whose research has been in the biology of medicine, and who now is president of Memorial Sloan-Kettering Cancer Center, in New York City, said in a recent conversation, "I'm not as fond of the notion of serendipity as I used to be. It seems to me now that if you get the research going—if you move away from speculating and making up hypotheses and actually go to work on the problem in the laboratory—then things are bound to begin happening if you've got your wits about you. You create the lucky accidents. Every now and then, something turns up in the course of exploration that's worth—as the guidebooks say about restaurants—a detour. I think that's when really important observations are made." Other scientists have emphasized this same point.

"And I think one way to tell when something important is going on is by *laughter,*" Thomas went on. "It seems to me that whenever I have been around a laboratory at a time when something very interesting has happened, it has at first seemed to be quite funny. There's laughter connected with the surprise—it *does* look funny. And whenever you can hear laughter, and somebody saying, 'But that's *preposterous!'*—you can tell that things are going well and that something probably worth looking at has begun to happen in the lab."

Luigi Galvani was an anatomist at the University of Bologna at the end of the eighteenth century. He was conducting experiments on nerves, using frogs. One day he threaded the legs of some frogs on copper wire and, by chance, hung them up in such a way that a circuit was made from the legs through the wire to the iron balustrade on which it was hung, and so to a part of the balustrade close to the frogs' legs. A puff of wind was enough to bring the legs close enough to the iron to complete the circuit: a spark snapped, and the legs jerked violently. (Still today, we speak of someone "galvanized into action.") It's been suggested that not Galvani but his wife first noticed the effect. Galvani pursued it with ingenious experiments, devising an arc of two metals to produce the contractions reliably, and published a first report in 1791. He had discovered electrical currents, in their first tame form—although, of course, it was later recognized that lightning, with which experiments had been started by Benjamin Franklin forty years earlier, was the same stuff in the wild. Galvani's discovery was as truly a matter of chance as any in science. But behold the power of patterns of thought: Galvani the anatomist insisted that the origin of the electricity was in the muscles and nerves of the frog. Alessandro Volta, a physicist at the University of Pavia, realized that not the animal tissue but the junction of two metals was producing the electricity—and upon that realization built, by 1800, the first battery.

Claude Bernard was a physiological investigator of high talent and—in the French intellectual tradition—even higher self-awareness; he founded and expounded experimental medicine. In 1850, Bernard surmised that the impulses that travel along nerve fibers induce chemical changes, which is true, and that these chemical changes should be detectable because they would produce heat. He cut the sympathetic nerve leading into one of a rabbit's ears, but left the other ear intact—expecting that the temperature of the first ear would drop compared to the other. Instead, the enervated ear grew warmer—by 4 to 6 degrees Celsius. "This increase can be appreciated very easily by the hand," Bernard said in his report that December to the Biology Society. Bernard had disconnected the blood vessels of the ear from the nerve that normally keeps them more or less tensed. Relaxed, the blood vessels dilated to allow more blood to flow through that ear, and faster. The ear was blushing. Bernard had discovered the vasopressor nervous system, which regulates blood pressure and the flow of blood through the flesh near the skin. Clarity of ideas and precision of technique can generate the exhilarating accident even when the intention at the start is contrary to the eventual result.

Or the original intention may be sensible, even correct, yet produce an unexpected observation so much more interesting that the research is lifted into a new trajectory. These are the times Thomas meant, when the actual work in the laboratory creates the lucky accidents. In 1889, Joseph von Mering and Oscar Minkowski, at the University of Strasburg, in Alsace, were studying the function of the digestive juices produced in the pancreas. They tried removing the pancreas from several dogs.

One of the animal keepers then noticed that the urine of these dogs attracted flies. But von Mering and Minkowski knew that flies swarm around the urine of diabetic patients. Indeed, that was one of the observations that had led to the knowledge, even in ancient medicine, that the urine of diabetics is sweet, and eventually to the discovery, in 1815, by Michel Chevreul, a French chemist, that it contains the sugar glucose. When the dogs' urine was analyzed, sure enough it contained glucose. Von Mering and Minkowski had produced diabetes experimentally. This was the first step—of many, and research still continues—to the identification of the causes of diabetes, the elucidation of the role of the hormone insulin in control of the levels of sugar in the bloodstream, and the development of ways to treat the disease.

Was the discovery chance? From one aspect, unquestionably. As is well known by now, the pancreas has within it small bodies of a different kind of tissue, which are called the islets of Langerhans, after the physician Paul Langerhans, who first discerned them. The pancreas discharges its digestive aids directly into the intestines. Insulin and other hormones made in the islets of Langerhans are discharged into the bloodstream. The functions of the two types of tissues—the pancreas itself and the islets within it—are thus entirely distinct. Yet from another aspect, since von Mering and Minkowski were well-trained physiologists embarked on an investigation of the pancreas, it seems inevitable that they would have made the discovery they did make. The chance that the laboratory assistant first brought their attention to the dogs' urine then appears to be mere detail.

By today's standards of purity in research, Pasteur, in his choice of problems, was crudely practical about the needs of commerce and manufacturing. The French wine industry in the mid-nineteenth century, introducing mass production for the mass market, was losing millions of gallons that turned sour. Pasteur had begun to elucidate the nature of fermentation in beer and wine; in 1865, he introduced the heat treatment of wine to stop fermentation and souring. Thus on a lake of bad wine he launched the science of bacteriology. The bacterial theory of disease grew from that, and led him to investigate a costly problem in animal husbandry, a disorder that kills fowl and is called cholera (though it is unrelated to human cholera). In the course of demonstrating that fowl cholera met the criteria for identifying a bacterial disease, one day in 1879 Pasteur or one of his assistants inoculated some chickens with a cholera culture from the wrong jar—it was old. The chickens got sick, but then recovered. Curiosity piqued, Pasteur inoculated the chickens again with a fresh culture known to be virulent. This time the chickens didn't even get sick.

Immunization was known in medicine; nearly a century earlier Edward Jenner had developed a vaccine for smallpox. But Jenner had worked entirely by trial and error; nothing was known about either the cause of smallpox or the physiological basis of immunity. Pasteur and those who had followed him into bacteriology knew, also, that bacterial strains vary in virulence. Confronted by the chickens that ought to have been dead, Pasteur made the leap that has become famous: that an attenuated strain of infectious agent may produce an illness from which an animal can recover with full immunity against the virulent form. In 1880, he announced a preventive treatment against fowl cholera—and in honor of Jenner called it a vaccine. From fowl cholera Pasteur went on to a vaccine against a bacterial scourge of sheep and cattle, anthrax. Only then did he turn to the discovery that made him the most celebrated Frenchman since Napoleon, the vaccination procedure against rabies. Modern immunization thus began with yet another kind of chance event that will be familiar in any laboratory: the productive careless mistake. Eighty years or so later, in recognition of the role that error often plays in the real work of discovery, and of the companion fact that error itself must be controlled if the interesting result is to be comprehensible, Max Delbrück, one of the founders of molecular biology, offered an ironical phrase for it: "the principle of limited sloppiness."

The tentacles of the sea anemone are tipped with an irritating poison, as any snorkeler who has brushed against one will know. In 1900, Charles Richet, a physiologist at the University of Paris, attempted to discover the lethal dose of this toxin by injecting laboratory dogs. As part of the series, he used animals that had survived a large dose to try the effects of a second injection several weeks later—and, as a matter of routine, he gave these animals doses at different strengths. To his astonishment, he found that animals given a second dose much smaller than the first—a tenth or even less—died in a few minutes. "At first I had enormous difficulty in believing this," he wrote later. That a few people react uncomfortably, even painfully, to small amounts of substances like ragweed pollen or strawberries or lobster, and so on, and indeed that grown men have been killed by a beesting, was generally known, though unexplained. "In spite of myself," Richet wrote, he had discovered induced sensitization, otherwise called anaphylaxis—the basis of all allergic reactions. After his initial incredulity, Richet saw what he had found. He coined the term and pursued the study to a Nobel Prize in 1913. His is a celebrated case of another kind of chance discovery that is familiar to almost anyone who has worked seriously in the laboratory: the times when what started as the control becomes the experiment.

In the decade beginning in 1895, physics was overturned and opened up by the extraordinary series of discoveries of new forms of radiation that began with the accidental observation of X rays.

The early part of that series is a chain of chance. In 1895, at the University of Würzburg, Wilhelm Conrad Röntgen was experimenting with the discharge of electricity through gases held in a glass tube at very low pressure. One evening he had nearby a screen coated with barium platinocyanide, which is a compound used to detect ultraviolet light because it absorbs the ultraviolet and emits the energy again as visible light—the phenomenon called fluorescence. While electricity was discharging in the tube, Röntgen noticed the screen brightly fluorescing. He put his hand between the tube and the screen and saw what no man had seen before, the shadow of bones in living flesh. He placed photographic plates, well wrapped, near the screen; they became fogged. A key placed on a box of plates left its shadow. He traced the effect to a spot on the wall of the tube where the glass glowed greenish, directly opposite the cathode, or negative terminal, of the electrical rig. He at once understood something of the potential of the discovery, tried all the obvious demonstrations, and, it's said, didn't even tell his wife what he was finding until he was ready to publish his first paper—with its famous photograph of the bones within the translucent flesh of a human hand. He named the new kind of radiation "X rays" for their mysterious properties. Within months, the paper was published in other languages, and physicians in many parts of the world were using X rays for setting broken bones and for internal examinations. In 1901, Röntgen was awarded the first Nobel Prize in physics for his discovery. But was the discovery chance? Once again, in immediate detail yes—and yet, Röntgen possessed that coated, fluorescent screen because he was interested in radiation of all sorts. Chance favors the prepared laboratory. Meanwhile, once Röntgen published, others who had experimented with electrical discharges in glass tubes recalled that they had had trouble with packages of photographic plates inexplicably fogged.

Chance continued to fluoresce spectacularly. In Paris, Henri Becquerel, whose father and grandfather had been eminent physicists before him, was aroused by Röntgen's announcement to wonder about the connection between X rays and fluorescence. His father, Alexandre Edmond Becquerel, had made a lifetime study of light. He had given special attention to the fluorescent compound potassium uranyl sulfate, which contains in every molecule an atom of uranium. Henri Becquerel, to see whether fluorescence contained any X rays in addition to the visible light, wrapped a stack of photographic plates securely in black paper, placed some potassium uranyl sulfate on top, and put the whole stack in the direct sunlight where the ultraviolet would excite the fluorescence. When he developed the plates, they were blackened. More trials showed that the radiation from the fluorescing compound would even penetrate aluminum or copper—something that only X rays were known to be able to do. But then the weather grew cloudy. Becquerel put a packet of plates away in a drawer—as it happened, with some traces of potassium uranyl sulfate lying on the packet. After several days, he decided to develop the stored plates, thinking that even without exposure to direct sun, some X rays might have been produced. The plates, developed, were heavily blackened. Neither sunlight nor fluorescence could be responsible. Experiments with other compounds quickly found that the uranium in the potassium uranyl sulfate was producing intense radiation. Becquerel had discovered natural radioactivity. Marie Curie soon afterward named it. Becquerel shared the Nobel Prize in physics in 1903 with Marie and Pierre Curie. The discovery fairly opened the revolution that overturned and rebuilt physics in the next twenty years. But was the discovery chance? In effect, Becquerel performed unintentionally a control experiment he should have designed. By the principle of limited sloppiness, the control became the experiment.

As anecdotes, all these instances of chance discovery have been recounted many times. Rarely reached are the questions they suggest about the interplay of incidental circumstances with the ripeness of problems in their time, about the relation of the individual to the scientific community, and about the kinds of preparation that sensitize the mind, or the patterns of ideas, known evidence, and preconceptions that facilitate or defeat the exploitation of chance. Certainly, the most famous chance discovery of them all was Alexander Fleming's observation, in 1928, of the bactericidal action of the mold *Penicillium notatum*. But this instance of chance discovery, too, raises questions—disturbing questions of a different kind, not how and why the discovery was made but why in the previous sixty years it was repeatedly prefigured without being followed up.

Fleming was a bacteriologist, and for most of a decade had been searching for antibacterial agents. In 1922, he had discovered an enzyme that is present in egg white, tears, and mucus and that chews through the cell walls of invading bacteria, causing them to burst. The word for bacterial bursting is "lysis," so Fleming called his enzyme "lysozyme." He recognized its remarkable powers of protecting the organism, but was unable to make a medicine of it. Fleming in 1928 was working at St. Mary's Hospital, near Paddington Station, in London. His laboratory was small and disorderly, papers and flat glass culture dishes everywhere. In warm weather, he left the window open. In 1928, he was writing an article on staphylococci, the germs that cause boils and infect wounds; in preparation, he grew the bacteria on gelatin in culture dishes, for examination beneath the microscope. Cultures sometimes got contaminated by molds, for a single airborne spore could start a colony. One day, talking to a colleague, Fleming noticed a moldy culture dish—and saw, too, that all around the colony of mold the clusters of staphylococci looked as though they had melted. They had faded and turned watery. The individual bacteria had burst. Fleming transferred a scrap of the mat of mold to a tube of broth. He also preserved the original culture dish.

Fleming's laboratory had few resources, but he was able to grow the mold, identify it as one of the *Penicillium* family, and prepare a crude extract. He tried the extract against bacteria, first in glass dishes, then by injection into animals, and showed that it was a powerful killer of several kinds of germs though not of some others. He named the active principle "penicillin," and published his first paper about it eight months after the chance observation. Further testing of penicillin's effects, and its isolation, turned out to be very hard. A decade later, the observation and the mold were exploited by a large group, directed by Ernst Chain and Howard Florey, at Oxford University; they isolated penicillin in 1939 and proved its therapeutic power and safety with human patients in 1941. The onset of war drove research into mass production of the substance to the United States, where it was achieved by the mid-forties. Fleming, Chain, and Florey shared the Nobel Prize in physiology or medicine in 1945.

Nobody supposes that Fleming spotted the mold's action or wrote his early papers with any knowledge that the discovery had been made several times before. Nonetheless, in the late nineteenth century biologists were fascinated by the notion that an intricate and merciless struggle for existence goes on throughout the living world. Many bacteriologists thought that different kinds of microorganisms might compete by inhibiting or poisoning one another's growth.

The English surgeon Joseph Lister—who in 1865 as a follower of Pasteur had introduced antiseptic procedures and sterile instruments into the operating theater—noted in his laboratory book, on November 25, 1871, that in a sample of urine containing bacteria, with some filaments of mold also present, the bacteria seemed unable to grow. He tried experiments to see whether the mold made the liquid an unfavorable medium for the bacteria, but these got nowhere. The mold, Lister thought, was one of the *Penicillium* family.

At that same time, John Tyndall, an English microscopist, was at work on an elaborate refutation of the ancient idea of the spontaneous generation of new life from nonliving matter. The point he was making can be seen in the title of the book he published in 1881: *Essays on the Floating Matter of the Air in Relation to Putrefaction and Infection.* In December of 1875, Tyndall noticed that a species of *Penicillium* that had floated into one of his experiments caused bacteria to burst; he communicated some of these observations to the Royal Society the next month, but did not pursue the effect.

In 1877, Louis Pasteur and a colleague, Jules Joubert, were growing anthrax germs in a flask of sterilized urine. They observed that some airborne microorganisms invading the culture could inhibit and even destroy the bacilli. The two tried experiments to see if the invaders, injected into guinea pigs that were inoculated with anthrax at the same time, would prevent the disease. The results were uncertain, and Pasteur dropped the work—though he reported it in a paper where, in conclusion, he wrote, "All these facts may legitimately encourage our greatest hopes from a therapeutic point of view." Then, in 1889, Paul Vuillemin, a French physiologist, first proposed the term "antibiosis."

These were but precursors. The discovery and thorough proof of the action of *Penicillium* against bacteria, and its potential for medicine, were made in 1896, by Ernest Augustin Clement Duchesne, a twenty-two-year-old student at the Medical Academy of the French Army, in Lyons. Duchesne, familiar of course with Pasteur's work, grew fascinated by the struggle for existence among microorganisms. He knew that mold cells and spores teem in the air, and will rapidly proliferate when they land on moist food; yet in water where bacteria were growing, molds were rarely to be found. He wondered whether the bacteria were inhibiting the growth of molds—and whether molds ever inhibited the growth of bacteria.

To qualify for the M.D. degree, Duchesne had to prepare research for a dissertation; he chose the antagonism between molds and microbes. As an experimental subject, he picked *Penicillium glaucum*, which grows the luxuriously furry, pale green colonies that everybody has seen on bread, jam, or rotting fruit. He grew the mold on some moist food. Then he added common intestinal bacteria, *Escherichia coli.* Several hours later, he examined the colony and found the bacteria dead. Pursuing that, he tried live animals. In the control, he inoculated guinea pigs with virulent cultures of bacteria deadly to them, *E. coli* or *Salmonella typhosa* (the cause of typhoid fever); in the main trial, he dosed other guinea pigs with material, drawn from the same virulent stocks of *E. coli* or *Salmonella*, to which a vigorous culture of *Penicillium* had earlier been added. All the animals that got only bacteria were dead twenty-four hours later; the animals that got *Penicillium*, as well, all survived. In his thesis, he called for the work to be repeated and extended, and wrote, "Further research may lead to new developments that could prove directly applicable to . . . therapy." The dissertation was published at the end of 1897; Duchesne got his degree, went directly into the army, and was assigned to a regiment of hussars stationed north of Paris. He did no more research; in 1902, he caught a disease of the lungs, apparently tuberculosis, that eventually forced him out of the army and that killed him in 1912. His thesis was ignored for fifty years.

Duchesne's case is tragic: not a chance discovery

but a chance failure to take up a discovery that was beautifully in hand. Why nobody went after it appears impossible to reconstruct. Gunther Stent, a molecular biologist at the University of California who has thought a lot about the history of science, suggested a while ago that some discoveries are "premature"—that they appear at a time when they can find no place in the prevailing ideas, habits of thought, and accepted kinds of evidence, and must wait for decades, for half a century even, to be rediscovered. He gave examples, though not Duchesne's. There can be no question that the intellectual era in which Duchesne devised his experiments should have been receptive to his results: he chose the line he did because everybody was thinking and writing about the struggle for existence and antagonisms between organisms. The prematurity lay elsewhere, in something simpler and more stringent: technology. If Duchesne had lived and had been free and financed, could he have taken the discovery much further? One feature clamors to be noticed in all the other instances where the antagonism of *Penicillium* to bacteria was recognized: again and again, the observation provoked a few crude experiments that turned out inconclusively and were abandoned. Fleming himself hardly got beyond the fact of penicillin's power—except that he named the stuff, a step that can be crucially important to gaining attention in science. Chain and Florey had the best colleagues, the most sophisticated equipment of the day, urgent support in time of war—and yet overcame great difficulty with not inconsiderable luck. For example, some common species of laboratory animals find penicillin poisonous, but Chain and Florey tested their first highly purified preparations in animals that tolerate penicillin as people do. And then the production of penicillin in quantity turned out to be by far the most complicated, costly manufacturing problem taken on up to that time by the American pharmaceutical industry; this, too, would not have been attempted without the pressure of war and the guarantee of government money.

The whims of chance are individual. Science is a collective enterprise to a degree that anecdote cannot convey. Penicillin's history warns that often the absolute limit is one of technique. Self-evidently, the most original scientific work at any time may require that the methods be invented new in parallel with the observations and the ideas.

Chance in the process of discovery is interesting but incidental to the things discovered. In a different sense, though, chance is fundamental to many of the phenomena that scientists want to understand. Chance being inherent in some of the processes of the universe, there are some things we can never know.

Scientists speak of such ignorance at two levels. First is classical ignorance. For Newtonian mechanics, the interactions of bodies small or large, from atoms and molecules to planets and stars, were strictly determined and therefore were—in principle—predictable. If you knew the total state of the universe at one instant, its state at any future instant was—in principle—deducible. In practice, such interactions often and quickly become so complex that they are genuinely and forever incalculable. For this reason, scientists often must reason from probabilities and by statistical inference. To predict the outcome of an election exactly and with absolute certainty is impossible. What if, on election day, a freak storm in part of the country keeps people away from the polls? More fundamentally, if the prediction becomes known before the election, what are the effects on people's intentions? But the standard example of classical ignorance is the total behavior of the individual particles in a waterfall: a problem that is strictly one of classical mechanics and is incalculably complicated, which is why the study of turbulence has been so difficult. Perhaps nothing else so sets off the daily habits of mind of scientists from other people's as does the willingness to accept the inevitability of uncertainty in some crucial matters, and the consequent necessity to act on the basis of good probabilities. Many psychiatrists smoke, but I know few biologists who do so.

Charles Darwin took the step that built chance into the foundations of modern biology. Unlike previous ideas of evolution, Darwin's theory of the origin of species by natural selection carried no promise of progress. Characteristics however useful that were not inborn to a creature but were acquired during its lifetime were not inherited by its offspring. Giraffes' necks had not got longer by generations of stretching: coldly, those individual ancestors of modern giraffes that were born lucky, with necks that grew longer than their brothers' and sisters', were more likely to reach enough food to survive to pass their lucky heredity to descendants.

Darwinian evolution is a two-stroke engine: hereditary variations among a creature's offspring provide the range of characteristics among which those best adapted to the immediate environment will be more likely to reproduce. The likelihood is statistical. Yet we now know that a surprisingly small increase in that likelihood, given enough generations, must have powerful effect in spreading a characteristic through an evolving population. Darwin's contemporaries relished the selective part of the process, the intensely competitive struggle for existence. Herbert Spencer, and not Darwin, introduced the slogan "survival of the fittest." Darwin himself saw the primacy of the other part of the process, providing variation to be selected from. He devoted four of the first five chapters of his great book to cataloguing and analyzing variability in plants and animals, domesticated and wild. Yet Darwin had an inadequate and mistaken idea of the mechanisms of heredity. Gregor Mendel, a Moravian abbot and naturalist, analyzed the statistics of inheritance of various characters in pea plants. His paper publishing the rules of heredity was contemporary to Darwin's *The Origin of Species*, but Mendel's work was not widely known or understood until the rules were rediscovered at the start of the twentieth century. Then Mendel's units of heredity were christened "genes"; and then Hugo de Vries, a Dutch botanist, introduced the crucial idea of mutation, or random change in the genes. Darwin's variability had found its primary mechanism.

Mutation is genuinely random: we now understand it originates when a location in a DNA molecule—any of a long line of similar locations that are, in effect, the letters in which the genetic message is spelled out—is slightly altered. The alteration changes one letter in the message, producing a new word there, or nonsense. The alteration can be caused by X rays or other radiation, or by chemicals taken in from the environment; where in the message the mutation strikes is arbitrary and independent of the message's original or altered meaning. Thus, the variations from which evolution has sprung originate in chance. They are not predictable—an instance of classical ignorance.

Still more fundamental is quantum ignorance. Out of the discovery of radioactivity grew a conception of the atom and events within it that culminated, in the late 1920s, in quantum mechanics. The equations of quantum mechanics accounted for the phenomena, but to make them work the idea of strict classical causality had to be abandoned at the subatomic level. In principle, to give a standard example, even though a sample of a given radioactive element emits radiation and decays at a characteristic and invariable rate, we cannot know at any instant for any given atom in the sample when it will decay. More generally, as quantum mechanics was being worked out, Werner Heisenberg introduced the uncertainty principle. He demonstrated that on the scale of events in the atom we cannot simultaneously determine both the position and the momentum of a particle, such as an electron or a neutron. There is a spot of uncertainty within which, if we find the particle's position more exactly, our knowledge of its momentum becomes less certain, while if we detect the momentum exactly, the particle's position becomes fuzzy. Bitterly attacked by Einstein among others, quantum randomness and the uncertainty principle have survived and are an inextricable part of present physics.

A Conversation with Manfred Eigen

"They called everything 'infinitely fast' that was faster than one-thousandth of a second. But I could easily estimate that the fastest possible reactions one could think of would require only one-billionth of a second."

Perhaps the most delightful kind of chance encounter occurs when two apparently unrelated problems meet and annihilate each other in the mind of the right person. In 1950, the unexplained fact that sound travels less well in the sea than in fresh water was brought to the notice of Manfred Eigen, a young physical chemist in Göttingen, who just then was baffled by a different problem, trying to measure reactions that were said to be immeasurably fast.

Eigen today is director of the Max Planck Institute for Biophysical Chemistry, in Göttingen. He is short and dapper, with a tanned, youthful face, white hair, and a compulsion to explain science. When he was a research student, in the early fifties, Eigen solved for his doctorate a problem in the mechanisms by which molecules or their components associate when they are in solution. This is the chemistry of electrolytes, meaning the compounds that break into ions, or charged fragments, in solution—as common salt, for example, dissociates into positive sodium ions and negative chloride ions. "That was looking at the chemistry of electrolytes more or less as a static picture," he said to me recently. "But then I became interested in the dynamics of these reactions—exactly how they progress physically to reach equilibrium. And how fast. And, naturally, I looked up in textbooks how to measure the speed of chemical reactions.

"I found in one textbook the remark that certain types of chemical reactions, say in the behavior of chemicals combining when they are in solution, are *infinitely* fast.

'Infinitely' means, of course, so fast that there is no hope ever to measure the speed. I looked up the time scale and thought a little about the chemical mechanisms; and I realized that they called everything 'infinitely fast' that was faster than, say, one-thousandth of a second. But I could easily estimate that the fastest possible reactions one could think of would require only one-*billionth* of a second. So here there was a time scale of six to nine orders of magnitude unexplored." One order of magnitude is a jump from one size or duration to the next one ten times larger, or ten times smaller. From one thousandth to one billionth—that is, from 0.001 to 0.000000001—spans six zeroes, six orders of magnitude, or, if one counts a billion not as Americans do, a thousand millions, but in the European way as a million millions, then the jump is, as Eigen said, nine orders of magnitude. The scale effects are severe. "A gap like this is a challenge to a scientist," Eigen said, with a cock of an eyebrow and a quick smile. "You say that perhaps, if you find the right trick, you could still measure those processes and fill up that gap."

This was Eigen's problem. At that same period, however, engineers were attempting to improve sonar, the method of detecting things underwater—the shoaling bottom, obstacles, schools of fish, submarines—by emitting pulses of sound and listening, electronically, for reflected waves. The duration of the round trip reveals the distance, supposing that one knows with sufficient precision how the water absorbs and transmits sound. "So this was the other thing," Eigen said. "People at that time were measuring the sound absorption of sea water. And they found that sea water has a very large absorption coefficient for sound waves. And physicists were wondering *why* sea water absorbed sound much more efficiently than ordinary water or distilled water. I thought I had an answer for this.

"Because I had investigated the properties of electrolyte solutions, I saw that this was a particular kind of mechanism. I suggested immediately that it couldn't be the sodium chloride in sea water, because my colleague Konrad Tamm had found that sodium-chloride solutions don't have large sound-absorption coefficients. But I suggested it must be the magnesium sulfate—another salt—which causes this. Sea water contains quite a lot of magnesium sulfate. And in solution the magnesium ions complex with sulfate ions; I had studied that and knew the mechanisms. So my proposal was that the energy of the sound was used to cause a dissociation of these complexes and an association of the ions—so that part of the energy was being put into chemical energy. And that was causing the loss, causing the sound absorption. We carried out experiments; and it turned out that magnesium and sulfate were the correct ions, and the answer."

Antiproblem met problem in Eigen's mental cloud chamber and the two vanished in a flash of understandings. "At that moment I saw *immediately* that this was the answer to the problem of measuring fast reactions. Because with sound, you can get frequencies of a thousand cycles, which you can hear, up to megacycles (millions of cycles) or hundreds of megacycles. And so with sound waves at the right frequency and energy, you can perturb the existing chemical equilibrium for the briefest instant—for that billionth of a second. And then see by simultaneous measurement of other parameters what difference the pulse of disturbance made in the physical chemistry."

The absorption of sound proved, indeed, to be a way to pulse energy very briefly into a chemical reaction under study, so that dissociation and reassociation, departure from and return to equilibrium, can be watched. These are called "relaxation techniques." The technology has grown complex, as Eigen once showed me on a walk through his laboratory in Göttingen, and it now uses, besides sound, pulses of intense light, pulses of radio microwaves, even the flash heating of almost microscopically small droplets of solutions. But in principle the idea is as simple as when Eigen first conceived it. For the invention of relaxation techniques for measuring ultrafast chemical reactions, Eigen shared the Nobel Prize in chemistry in 1967 with George Porter and Ronald G. W. Norrish, two English chemists who had independently explored some related methods.

A Conversation with Jocelyn Bell Burnell

How pulsars were first observed: "The discovery was almost totally unexpected."

In England during the Second World War, when radar was just being introduced, physicists working under intense pressure on the improvement of the technology were sometimes told by operators manning radar equipment that they were getting signals that were not reflected from aircraft, were not the massive radio noise from the sun, but were coming from empty stretches of sky—apparently from the reaches of outer space. After the urgency of war was over, one of these physicists, Martin Ryle, came to the Cavendish Laboratory, at the University of Cambridge, to try to find the source of those signals. In an empty field down a side road just out of town, Ryle set up a simple antenna of his own design—the first radio telescope. He quickly found that there were radio signals at all sorts of wavelengths and all sorts of strengths coming from space. Some of these could be identified with particular stars known from conventional optical astronomy, or with known interstellar clouds of gas or known nebulae, but others had no visible source. An entire new dimension of astronomical observation was opening up, far out beyond the long-wave, red end of the visible spectrum. Other people had noticed radio signals from outer space, too; in fact, there had been reports from ham radio operators in the United States as early as the 1920s. Several other centers of radio astronomy also were started, soon after the war; the one at Cambridge remained among the leaders, attracting excellent physicists and astronomers—notably Anthony Hewish—and many able students, and attracting, as well, the financial support that enabled the building in those empty fields of arrays of ever larger and more sophisticated antennas.

Jocelyn Bell (she is now Jocelyn Bell Burnell) came to

Cambridge as a graduate student in 1965. For as far back as she can remember she had wanted to be an astronomer. She was born in Northern Ireland, where her father was an architect. Her father, in fact, designed an observatory and planetarium in the town of Armagh, and as a child Jocelyn sometimes went with him when he had work to do at the observatory. "The astronomers there were very good to me, when they knew I was interested," she said not long ago. "They gave me a lot of advice. One bit of advice was that if you want to be good at astronomy you've got to be able to stay up late at night. And even as a teenager I knew that I wasn't able to stay up at night, so I went home that day very depressed.

"I found out fairly soon after that that there was a new branch of astronomy, called radio astronomy, and you didn't normally have to be able to stay up late at night to do that," Burnell went on. "So when I went to Cambridge, to do radio astronomy, I chose a project that I was fairly certain would not involve me staying up late at night. Because it was based on something about the sun that Professor Hewish—he was my supervisor—had discovered a few years earlier. Which was that when radio waves from a distant star or galaxy come to us here on earth, they come through the space between the planets—and that space is not quite empty, in fact. It's filled with what's called the solar wind. The extreme outer surface of the sun is blowing off all the time and spills right out into the solar system. And it's not a completely smooth wind. It's slightly patchy or blotchy—it differs in density. And as the radio waves come to earth through this slightly blotchy solar wind, they get bent slightly. So if your radio telescope is observing a distant radio source that's very compact—with a very small apparent diameter, seen from earth—the source appears to twinkle. But if you're looking at a radio source that's fairly broad, extended, it doesn't twinkle. There's an analogous effect in optical astronomy: one way that most people know to distinguish between planets and stars is that stars will twinkle whereas the planets give a steadier light. And that's because a planet, though intrinsically smaller than a star, is so much nearer that it appears more extended—so where, in this case, the variations in density in our own atmosphere make the light from the star twinkle, the more extended source does not twinkle.

"So this offered a way to pick out the compact radio sources and distinguish them from the more extended ones." A couple of years before Jocelyn Bell had gone up to Cambridge, radio astronomers—including Ryle and Hewish—had been shaken by the discovery of certain outlandish radio sources—extremely compact, yet broadcasting hundreds or thousands of times more brightly than any other sources including entire nearby galaxies, and, despite all that, apparently among the most distant objects yet observed in the universe. Astronomers called them "quasars," a word collapsed from "quasi-stellar radio sources." They were among the most fascinating objects in the sky. Hewish's idea was to do a radio survey of the sky, in order to catalogue quasars by looking for sources that twinkled. But radio sources don't usually twinkle by night, because one is looking away from the sun and the densest portion of the solar wind. "So I had—I thought—a daytime type of astronomy," she said. "So we built ourselves this enormous antenna system. It covered about four and a half acres of ground. And we sledgehammered wooden posts into the ground, about a thousand posts, and strung the actual antennas, copper wires, between the posts, and laid cables connecting them all together. I think it was 120 miles of wire and cable. It took a couple of years to build.

"And when it was built, then it was my job to start operating it, and to analyze the data that came pouring out. They came out on a strip of paper, a long sheet of blue-squared paper that the tracing pen moved over. We got a hundred feet of chart paper every day. I operated the telescope for six months, which meant that there was three and a half miles of chart recording by the end.

Anybody else would have got the computer to analyze it, but Muggins"—she gestured to herself—"sat down to analyze it by hand. Well, perhaps you wouldn't give it to the computer straight away, because with new equipment you want to get the feel of how it's performing and check that all's going okay.

"Fairly quickly you got used to picking out the twinkling sources. But unfortunately, the other thing that the radio astronomer gets is man-made interference. Radio telescopes are very sensitive. They have to be to pick up cosmic signals, which are very weak, coming from far across space. The energy you use to turn a single page of a book is more than all the radio telescopes have collected since the beginning of radio astronomy. The signals need amplification by factors of ten to the power of thirty before they can be recorded and studied. So almost anything in the vicinity that produces electrical interference will be picked up by the radio telescope. You get problems with badly suppressed motor-car ignitions. Thermostats kicking on and off. Arc welders. If the rules reserving frequencies to radio astronomers haven't been strictly observed, you pick up taxi radios, airplane altimeters, police-car radios. But fairly quickly you get used to recognizing the characteristics of the twinkling sources and the man-made interference.

"After a bit of chart analysis, I noticed there was something slightly peculiar on the record, late at night, from one particular patch of sky. It didn't look exactly like a twinkling source, and anyway twinkling sources shouldn't twinkle in the middle of the night. They only twinkle when they're in the daytime sky. And it didn't look exactly like man-made interference, either. It occurred a number of times when we were looking at that patch of sky. I discussed it with Tony Hewish, and we agreed that perhaps it deserved a closer look. It might have been some kind of flare on a star, where it temporarily becomes much brighter and then fades again. Or it might be a point source. One thing we were trying to do, as well as detecting the twinkling sources, was measure their angular diameter. How broad across were they? And you do that by studying the amount of scintillation—of twinkling. But it was difficult to calibrate that. And you remember, of course, that a point has zero radius. So we wondered if we had stumbled across some kind of source that was virtually a point source—because that would have helped with calibration a lot.

"The way you get a closer look is much the same way you get a photographic enlargement. You run your chart paper under the pen much faster than it normally runs, and everything becomes more spread out and enlarged and you can see it. As the student, it was my job to go out to the observatory each day at the time that source was passing through the telescope beam, and to switch the chart recorder onto a high speed. I did this very conscientiously for several weeks. No sign of anything, just the usual receiver noise. And garbage. One day, in disgust, I deliberately skipped it and went to a rather interesting lecture. Next day, on my recording there was this funny little bit of scruff again. So it hadn't gone and died on me. The next day I went out again to the observatory, all diligent and enthusiastic, and ran the fast recording, and it came through!

"It was this series of blips on the recording. And they were equally spaced, about one and a third seconds. Which is a suspicious frequency. If somebody had been playing around with a signal generator in one of the labs nearby, that's just the sort of frequency they might have set their signal generator to. I telephoned Tony Hewish. 'Oh, well, that settles it,' he said. 'It must be man-made.'

"Even so he came out to the observatory the next day, at the right time, and bless its little heart, once again the thing produced a series of pulses. Flashes. In retrospect, I appreciate how lucky I was, because we all know these things rarely perform to order. Just when you want them to put on a good display, they never do. But it was there.

"And that's where the problem started, to be honest.

We weren't leaping up and down shouting 'Eureka!' We were really rather worried. We had never seen anything like this before from the sky; we didn't expect to see anything like this from the sky. We couldn't believe it could be a star, because it was flashing at the rather rapid rate of one and a third seconds. Stars are big cumbersome things, when you get to them. We knew of stars that flash at periods of some *hours.*

"The precision and rapidity—one and a third seconds—is suspiciously man-made. If it's going to flash at that kind of rate, you expect the object to be something much, much smaller. More nippy, more agile, to flash at that kind of rate. We tried to explain it as being something man-made, and this didn't fit very well because it was going around with the stars. It kept the same place in the sky all the time. Whereas if it had been somebody on earth, he would have been operating to earth time, where the day is twenty-four hours—while in star time the day is twenty-three hours and fifty-six minutes." The four-minute difference comes from the fact that the earth, turning on its axis once a day, is also going around the sun—but not around the stars. "So if it was Joe Bloggs going home from work in a badly suppressed car, he was getting home about four minutes earlier, each night.

"So it couldn't be a star. And it wasn't man-made interference. But it went round the sky with the stars. So it *had* to be a star—or something out there. We had a couple of months of agony. We had to think of ways the *equipment* could be causing this. One of the nicest things was when another telescope at the observatory picked it up. That showed it was not something wrong with the antenna and receiver. But was it somebody's radar signals bouncing off the moon? Was it a satellite in a funny sort of orbit, going *bleep-bleep-bleep*? That didn't make sense either.

"One of the ideas we facetiously entertained was that it might be little green men—a civilization outside in space somewhere trying to communicate with us. Now, radio astronomers don't really want to believe that little green men will contact us. There may well be other civilizations out there in space, but I think the chances of making contact are pretty small. So it did seem to be a little bit going off the deep end to say we had detected signals from another civilization. But the name stuck. We referred to this radio source as LGM-1.

"Just before Christmas, 1967, things began to get very sticky. I was several thousand feet behind with the analysis. The evening before I went off for Christmas, just before the lab shut for the night and I would be locked in, I was analyzing a bit of chart from yet another part of sky. When there seemed to be on it something that looked just a little bit familiar—this same sort of scruff! Not exactly a twinkling source, not exactly man-made interference. I searched back through other records of the same patch of sky and, sure enough, it did seem to occur on a number of occasions. We estimated the distance of the source, and reckoned rather crudely that although it was well outside the solar system, it was well within our galaxy.

"One of the tests that Tony Hewish initiated stemmed straight from this idea of it being little green men signaling to us. He argued that they would be on a planet going around their sun, just like us on earth going around our sun. And that as their planet moved around their sun, some of the time it would be coming towards us, and some of the time moving away from us. And when it's coming towards us, the flashes would tend to get piled up on top of each other and the gap between them would be shorter, and when it's moving away, the flashes would get stretched out and the timing would be slightly longer. So he started some very accurate timing checks to try to detect this change as their planet went around their sun. In fact, he only succeeded in showing that the earth goes around our sun, no sign of any motion of the source. But that was useful, and it also established that whatever it was was fiendishly accurate—comparable with some of the best clocks man can make. Which implied it had to have great reserves of energy. So that it could keep on flashing dead accurately without any serious depletion of energy. Without any serious slowing up.

"Well, the discovery of the second one was quite a relief in one way because it scotched the little green men idea. You don't really expect to have two lots of little green men, both choosing to signal inconspicuous planet Earth, both choosing the same rather unspecial radio frequency.

"Then one afternoon after Christmas, on a chart from yet another part of the sky, there were *two* lots of this scruff! Slightly different times—an hour or so apart. I did begin to wonder if I was seeing things—had too good a Christmas or something. But later in January we confirmed those two as well. So we then had four of these things from totally different parts of the sky. It definitely couldn't be little green men. It really began to look much more as if it was some kind of star. Then somebody in the group remembered reading a paper about the vibrations of a hypothetical type of star called a neutron star. Way back in the 1930s, there were some theoreticians—with a great elasticity of mind, we felt—who had been studying different states of matter. They reckoned that there could be another stable state of matter so dense that a cubic centimeter, the volume of the tip of your thumb, would weigh a hundred million tons. They reasoned that if you took a star, and somehow forced it to shrink down to this density, it might come to be made up largely of neutrons. And a neutron star would weigh as much as an ordinary star like the sun, but would be only about ten miles across.

"It appeared that these neutron stars could well satisfy the criteria we had found. They were massive and they were compact. So that they could flash with great accuracy and great rapidity. We didn't understand how they would produce a flash of radio waves once every second.

We still don't. It's something to do with the neutron star spinning. There is perhaps something on the surface, a hot spot, a bright spot. And each time that comes around we get a flash. A bit like a lighthouse—though I don't want to pursue the analogy too far.

"We first got the pulses at the end of November of 1967, and sent a paper off in early February of the following year. When we published, we weren't sure between neutron stars and white dwarfs, the next biggest possibility—and already known to exist for sure. We stayed firmly on the fence. Six months later, the scientific community thought it was neutron stars.

"We still don't know very much about neutron stars, but we know enough not to go near them even if we could. These neutron stars must have a very strong magnetic field, about a million, million gauss. The earth's magnetic field is about a half or a third of a gauss. The whole thing is spinning—in the case of one pulsar in the Crab Nebula *thirty times a second*—like a massive generator.

"So they've got rather a lot of gravity and a strong *gradient* of gravity. Which means that the pull on your feet is a lot stronger than the pull on your head. We suspect that neutron stars have a solid crust, half a mile or so thick. Largely iron nuclei. Immediately inside the crust, you come to a region of more conventional elements that contain an excessive number of neutrons. Inside that material there's neutron superfluid. A superfluid is one with no viscosity, no resistance to flow. If you put it in a beaker, the material would flow up the sides and out over the edges. It's got no electrical resistivity, either. You can send a current through it, and the current will never die. And inside that, there is probably a core. But the very center of a neutron star is getting really hypothetical. We still don't know how material behaves when you pack things that close.

"Mathematically, neutron stars are very difficult to handle, partly because of the speeds. You're dealing with rotation at speeds near the velocity of light. You have to take into account relativistic effects. Very messy. And the fields—gravitational, magnetic—are so great that the atoms are no longer spherical but cylindrical! What we need is a way of *modeling* this ghastly mix of charged particles and magnetic fields and rotation. To the pure physicist, neutron stars are interesting because they show material at densities that we could not conceivably achieve here on earth. If we want to understand the basis of nature, how nature works when it's pushed to its limits, we must look out into space.

"The discovery was almost totally unexpected. We learned later of a radio astronomer at another observatory—I won't say who or where—who several years earlier was observing a portion of the sky to the right of Orion, northward, where we now know there to be a pulsar. And he saw his pen begin to jiggle. And he was about to go home for the day, and thought his equipment was misbehaving. And he kicked the table, and the pen stopped jiggling."

5

Feedback

Imagine that you and some friends, with small children, are picnicking on a grassy, mossy sward at the top of a cliff with a fine view of the sea. The precipice falls away sharply and is not fenced off, so you have set the hamper down at a level spot about ten feet back from the edge and are spreading out the lunch—when voices make you look up. You see two of the children playing with a stick very close to the edge. The stick bounces to the edge and over. One child yells and rushes to the brink to look. Instantly, involuntarily, your legs and stomach muscles tense and you flinch backward from the cliff edge as though you yourself were in danger—as though willing the child to pull back. Immediately, then, and with an intake of breath, you shift toward the child, reaching out an arm. But at once you think that you might startle the child, so you don't call out sharply and you don't jump to grab hold. Instead, you say, quietly, something simple like "Sandwiches are ready—come and eat."

We'll take that sequence of reactions apart in a moment.

Our understanding of the way things are interweaves great recurring themes, plays upon certain fundamental and all-informing ideas. The number of the themes of science is surprisingly small. Most have been developing for a long time in the background music of the world-as-we-know-it. One of these themes, indeed—and I'd say the most powerful of all, except that to rank them is senseless—has a lineage that stretches to antiquity. Atomic theory—the idea that all the complexity and variety of the world and the universe is built up from a few kinds of unimaginably small particles that are vastly multiplied, endlessly combined and permuted—took its modern form only in the nineteenth century, to be sure, but was first proposed twenty-four hundred years ago by Democritus, in northern Greece, and first popularized by Lucretius, in Rome several centuries later. Nearly as old is the simple but audacious idea that the earth is round. Some of the great themes were sounded at the start of the scientific revolution, in the sixteenth and seventeenth centuries: the heliocentric solar system, with Copernicus; modern mechanics, with the physics of Galileo; the circulation of the blood, with William Harvey; the calculus, with Leibniz and Newton. New observational realms were called into being then, too. The early telescopes and microscopes opened much more than the invisibly far and the invisibly small; they opened for the first time the possibility that different things may be going on, astonishing things, where our unaided senses cannot reach. The nineteenth century introduced three themes of shattering consequence: the cell in biology, natural selection in the evolution of species, and the unity of electrical and magnetic phenomena. Then the start of the twentieth century—in a burst of less than six years—saw the founding of genetics with the rediscovery of Mendel's laws (but this genetics was itself an atomized mechanism of heredity), the founding of quantum theory by Max Planck and Albert Einstein, and the founding of relativity.

The phenomenon called feedback appears often in nonliving processes and at every point and every moment in the processes of life and of thought. The patterns of feedback have subtle explanatory power. They bring together matters startlingly divergent—the biochemistry of the cell, for example, the engineering of computers, the price of beef, the way children learn to talk. Indeed, the idea of feedback interlocks with other ideas, about the nature of information and messages, to form the set of theories from which has sprung the second industrial revolution. For all that, the idea of feedback is simple. It is far more obvious, once stated, than almost any other of the chief themes of scientific understanding. The term itself is well naturalized in common speech—though it's sometimes used sloppily, as a mere synonym for "response." Yet feedback is the infant among the great themes. Feedback and the associated concepts of information theory emerged explicitly only at the middle of this century. Its history has not been written.

The child gambols on the brink of disaster and you knot up and cringe: your imagination is a half-step ahead of the child on its current trajectory and the picture is so vivid, the outcome so vital, that your body twitches in completion of the act that you'd perform yourself if you were where the child is. You recoil from the edge. But immediately, of course, you start to act in order to change the imagined future—you draw breath to shout, put an arm out to grab. Then your imagination, still ahead, sees the child startled into worse peril. Once more you change, suppress the response; once more you imagine an act of yours with a future, this time the children lured not dragged away from danger. Each correction is made in relation to a mental picture and generates the next move and the next mental picture. The all but instantaneous to-and-fro, in a closed loop, of act and consequence, corrected act with new projected consequences, further corrected act—this is the pattern that is feedback.

Feedback was first deliberately embodied in mechanical devices for controlling rates of flow of liquids or rates of motion of machinery. Precursors of such devices can be found in the engineering of antiquity, Hellenistic and Chinese, but the principle was never sorted out and all such applications were forgotten, at least in the West, by the late Middle Ages. In the eighteenth century, what we would now call feedback controls began to appear for regulating the temperature of industrial furnaces and the action of water mills and windmills. These remained isolated inventions, intuitively arrived at, until 1788. Late in that year, the Scottish inventor and builder of steam engines James Watt—for whom the unit of energy is named—designed a device for regulating automatically the output of his steam engines and had the factory construct the first one. He called the device a "governor," and didn't bother to patent it. A steam engine—driving rows of heavy mechanical looms, for example, in a cotton mill of nineteenth-century Lancashire—is subject to slow fluctuations in the temperature of the steam and to sudden variations in work load. The governor was an upright shaft, belt-driven to spin with the speed of the engine. Two arms were hinged to the shaft, near the top, and at the lower end of each arm was a small heavy ball. The device was fixed at the valve that supplied high-pressure steam to the piston cylinder. When the speed of the engine increased, the governor spun faster and the weighted arms flew up like a dancer's skirts. The rising arms steadily closed the valve, throttling the pressure of steam supplied to the piston cylinder; when speed decreased, the arms fell so that the governor eased the valve open again. This is feedback control. Flyball governors are still used today.

As the industrial revolution flourished, other inventors designed variable brakes and moderators to regulate other machines by using changes in output to adjust the input. Some were exquisitely precise: one steadied the drive that pointed the telescope at the Greenwich Observatory, down the Thames from London. Curiously, in that inquisitive, inventive age the principle behind these contrivances got little attention. Perhaps they seemed too simply mechanical to merit study. On March 5, 1868, though, a brief paper, "On Governors," was read before the Royal Society of London by James Clerk Maxwell, a Scottish mathematician and physicist of great penetration and range who had recently been made the first Cavendish Professor of Experimental Physics at the University of Cambridge. Maxwell pointed out that the motion a governor governs can be analyzed into two components, namely, a steady motion plus a disturbance—the disturbance being typically an oscillation that the governor is supposed to damp down. He put this analysis into mathematical form. He saw, further, that in some circumstances the action of a governor will not stabilize the motion it is supposed to regulate. "If, by altering the adjustments of the machine, its governing power is continually increased, there is generally a limit at which the disturbance, instead of subsiding more rapidly, becomes an oscillating and jerking motion, increasing in violence until it reaches the limit of action of the governor." In other words, if the governor overcompensates it drives the entire system into wilder and wilder swings. The transition is abrupt, and corresponds to a change in the signs of factors in the equations from negative to positive. Maxwell's observations remain fundamental to this day; but he said he was able only partly to solve the equations of stability.

In the 1920s, the principle of feedback showed up in an entirely different context—as the fundamental pattern in the approach to psychology formulated by George Herbert Mead, an American behaviorist. Mead was a colleague of John Dewey, at the University of Chicago. Unlike Dewey, whose views, particularly about education, were widely publicized and influential, Mead published little. His ideas spread through his lectures and teaching. Mead was fascinated by the problem of language—of how it functions as the very fabric of human interaction, and of how children learn to use it. He observed that at the earliest stage a child, obviously, makes a great variety of speech sounds, some of them imitative of what he hears, but that the parent responds only to those sounds that seem closest to actual words. In this way, the parent selects and reinforces those of the child's acts that seem correct. That much is orthodox, simple behaviorism. Mead went a crucial step beyond simple behaviorism. He put into the very learning of speech the element of imagination. He said that the word not merely signifies the act, it *is* the act—but the act enormously foreshortened. From the earliest learning of a word, it takes its meaning, he said, in "the imagined completion of the act." You who say the words, and I who hear them, each in imagination completes the acts of which the words are the first incipient motions. And each of us observes the other constantly—the other's gestures at the moment, and the words said in continuation or response in the immediate next moment—to verify that both of us have completed the act, in imagination, much the same way. But suddenly I look puzzled and more alert. At that tiny cue, you hear a second time, in the mind's ear, the words you just used and realize that although you meant them one way (though you imagined the act completed in one way), they could also be taken another way. My odd glance makes you follow again the flight of the arrow of the words—but now in the direction that, suddenly, you suppose I must have taken them. And you say, "No, no, I didn't mean *that*. What I meant was—" And you repeat yourself in words different enough so that your imagined completion still follows—but not my incorrect one.

In this way, ceaselessly, we check and adjust our use of language. We do it automatically, and all but unconsciously unless the misunderstanding is conspicuous. Would you like to demonstrate that? In a telephone conversation, the visible cues and gestures are cut out, yet the responses are otherwise as immediate as ever. Next time a friend is recounting something to you on the telephone in that sort of conversation where one person talks at length and the other is reduced to the occasional "yes" and "oh" and "um-hmm"—try silence. Say nothing. Withhold those signals, the *yes*es and *um-hmm*s. The effect is almost cruelly disconcerting. The other's flow of talk, within twenty seconds or so, falters, stumbles, dries; then comes the question, "Are you still there?"

Of course, what we do without cease with each other, as adults, checking and confirming understanding, we redouble with small children. The adults around a child work continually with words to bring him into their society—which means to train him in their particular ways of language. Mead suggested that the child in his struggle to understand and to be understood develops, of necessity, a sense of the "other" he is talking with and whose imagination—simultaneously and in parallel with the child's own—is completing the acts indicated by the words. As the sense of the other develops, then only does the child's awareness and definition of the *self* also develop. From the beginning, one's identity is formed and focused by language.

At a crucial point, Mead said, the notion of the "other" that the child has developed from moment to moment in individual interactions becomes something more permanent and more powerful—what is called "the generalized other." Mead's term is awkward and not widely used; the generalized other is the social group as held in the mind as though for reference. In particular, it is the facility—so vital, apparently so simple—to hear oneself speak. Part of the mind takes on the task of following out the course of one's words as though they were someone else's, to make sure, instant after instant, that they are leading where they were intended to lead. Adult language is impossible without that facility. Even what seems like the slightest impairment will destroy speech. In one experiment, a psychologist rigs up a set of earphones and a tape recorder so that the words you are speaking are repeated at you with a delay of one-fifth of a second—and the effect is catastrophic, making it impossible to complete even a single long sentence. Those who write must be especially responsive to the generalized other. Without the help of moment-to-moment response from a listener, the writer must perceive in his own words the ambiguities that will confuse or distract the reader, and must puzzle out how to suppress them. Experienced writers talk of their "ear." Sometimes they will stop and read a passage aloud to themselves. Robert Graves, poet and novelist, found the phrase for what the writer must cultivate: "the reader over your shoulder."

With language, man binds time. Language, with this facility for monitoring it and adjusting it, is the tool we use to predict the future on the basis of the past—to predict alternative futures that may result from alternative present acts, from choices. In a lecture in 1924, Mead defined "value." The definition is perhaps his most compressed and brain-teasing statement of the phenomenon we now call feedback. "Value: the future character of the object insofar as it determines your action to it."

It has become a commonplace that the first industrial revolution, that of the eighteenth and nineteenth centuries, replaced muscle power with machines to perform repetitive physical labor while the second industrial revolution, today's, is replacing brain power, used to perform repetitive mental acts, with computers. The term "feedback" itself first appeared in print in 1920, applied to early radio circuits. In the late thirties, big calculating machines began to be developed intensively. Although their physical design was still limited to arrays of mechanical relays (later replaced by vacuum tubes and then, of course, by transistors), the theory of high-speed computing quickly took its present fundamental form. In 1940, with Europe at war and the United States arming, the mathematician Norbert Wiener, at the Massachusetts Institute of Technology, prepared a report that listed certain essentials found in all computers developed since. These included having the mathematical logic built into the machine; binary rather than decimal arithmetic; electronic rather than mechanical operation; copious, flexible data storage.

Wiener worked no further with computers himself, just then. His main project early in the war was the improvement of antiaircraft gunnery.

The problem was aiming. Airplanes were so fast that all necessary computations had to be built into the guns' controls. Indeed, Wiener wrote after the war, unlike any previous target "an airplane has a velocity which is a very appreciable part of the velocity of the missile used to bring it down." The shell had to be fired so that it and the target would meet at one spot in space at a moment in the future. Predicting that spot and moment was made even harder by the fact that the pilot can take evasive action—though within limits set by the controls, by the strength of the aircraft, and by the stress that the pilot can endure without being knocked unconscious. The problem, therefore, was to predict the position of an object traveling fast on an irregular curve.

"To predict the future of a curve is to carry out a certain operation on its past," Wiener wrote. He did not know of Mead—yet the similarity of their thinking, even of their statements, is inescapable. The parallel that Wiener did perceive—that he was uniquely placed to seize upon—was between the purposes of computers and of control devices. "It will be seen that for the second time I had become engaged in the study of a mechanico-electrical system which was designed to usurp a specifically human function—in the first case, the execution of a complicated pattern of computation, and in the second, the forecasting of the future."

In that context, Wiener could perceive that the problem of designing a control for antiaircraft guns called for an answer to the question: How do humans forecast the position of their bodies in relation to other objects, in order to perform even the simplest acts? Thus the similarity between the control processes of machines and of minds was recognized for the first time. Wiener identified a concept crucial to both of the two industrial revolutions. "I came to the conclusion that an extremely important factor in voluntary activity is what the control engineers term *feedback*," he wrote in 1948. He explained it: "When we desire a motion to follow a given pattern the difference between this pattern and the actually performed motion is used as a new input to cause the part regulated to move in such a way as to bring its motion closer to that given by the pattern."

Actual performance is continually compared to desired performance and the difference used to correct performance. Simple but subtler than it seems: remember learning to catch a ball, or the first time you turned a corner on a bicycle. Wiener offered as the best mechanical example a form of steering engine used in ships. The engine, deep in the ship where it is connected to the tiller, has two valves, arranged so that if the tiller is too far to the left, one valve is eased open and the engine pushes at the tiller, from the left, until the valve closes, while if the tiller is too far to the right, the other valve opens and the engine pushes from the right. Up in the wheelhouse, the steersman turns the wheel, which shifts the control apparatus to right or left in relation to the tiller, opening one valve until the engine, by shifting the tiller to the new correct position, closes the valve again. Wiener noted, just as Maxwell had, that overgoverning will cause the system to fail: "A feedback that is too brusque will make the rudder overshoot, and will be followed by a feedback in the other direction, which makes the rudder overshoot still more, until the steering mechanism goes into a wild oscillation or *hunting*, and breaks down completely."

For the comparison to human action, Wiener proposed the simplest possible thought experiment: What happens when you pick up a pencil? Nobody, not even the trained anatomist, consciously, muscle by muscle, decides upon the series of contractions necessary to reach out arm and hand, thumb and forefinger, to seize the pencil. Rather, what one wills is the aim—to pick the pencil up. Then one acts so that, from moment to moment, the gap by which the pencil is not yet picked up is narrowed. But for that to be possible the gap, the amount by which one has as yet failed to pick up the pencil, must continually be reported back through the nervous system. The report will normally be visual in part, of course, but it comes chiefly through the kinesthetic sense—one's skeletal, muscular sense of the positions of one's body and limbs.

If the steering engine, operating by feedback, is truly a model or analogue of acts like picking up a pencil, when neurological feedback is interfered with, nervous disorders should result. A class of disorders exists, Wiener knew, called ataxia, in which the patient cannot perform simple motor acts. One form of ataxia is a classic sign of syphilis in its advanced stages. A standard test for the condition is to ask the patient to shut his eyes and then touch the tip of his forefinger to the tip of his nose. Advanced syphilitics miss. They can't tell, with their eyes closed, where their finger and nose are. Their kinesthetic sense is impaired or missing. They can't walk with their eyes closed, either. Autopsies find that the disease has attacked the central nervous system, interrupting the transmission of the kinesthetic information.

But the steering engine as a model of human acts suggested strongly that other pathologies should exist—those in which feedback overshoots. Wiener knew of nothing like that, so he went to an associate who was a physician to ask if there was any nervous disorder in which the patient, trying to do something like picking up a pencil, overreaches and goes into uncontrollable oscillation, wild hunting swings. He got the prompt reply that such a disorder was well known to medicine. It is called purpose tremor and is often found in cases of damage to the cerebellum, the part of the brain that coordinates movement and maintains muscle tone.

The living organism as a machine, man as a machine, has furnished a topic for philosophical speculation for centuries, since Descartes and before. Feedback gave the comparison its first real content. Early in 1943, with two colleagues, Wiener published a paper to draw these parallels and to talk about feedback in human behavior. This first try was inconclusive, a rambling philosophical statement in need of mathematical stiffening. But Wiener found the similar role of feedback in the behavior of man and machine hugely stimulating. Above all, feedback made sense of the idea of purpose in human behavior. It reconciled determinism and choice.

At about that same period, Wiener and his associates began to see another, still deeper set of parallels. Their problem in the design of controls turned out to be inseparable from certain other problems, which communications engineers were encountering when they tried to imagine ways to transmit information reliably at high speeds. At the base of both control and communication lay the "fundamental notion of the message," however it might be transmitted, electrically, mechanically, or nervously. What the feedback loop carried was information. What a computer processed was information. What a telephone line carried was information; what a radio or television broadcast carried was information. And high-speed reliable transmission of information posed a dilemma.* On one horn of the dilemma, to pack the most information into the message requires that it

---

*It also posed the need for definitions of information and of the amount of information. Wiener and others, notably Claude Shannon, at Bell Telephone Laboratories, arrived at the same definitions almost simultaneously. Information was to be counted in terms of the least and simplest unit, which is the transmission of a single decision between equally probable alternatives—that is, a heads-or-tails, a yes-or-no, an on-or-off. The unit of information was naturally and directly expressed in the binary notation, which was coming to be used in computers, where the digits are 1 and 0 (so that "2" in the decimal system is written 10 in binary notation, "3" is written 11, "4" is written 100, "5" is 101, and so on). In binary notation, 1 can stand for "yes" and be carried as the *on* position of a computer switch, relay, vacuum tube, or transistor; 0 can be "no" and *off*. Thus, the smallest unit of

contain as little repetition—as little of what the engineers call redundancy—as possible. To save sending redundant information—and the savings could be enormous: doubling and redoubling, for example, the capacity of a long-distance telephone or television cable—the message could be analyzed at the transmitter to find and trim out any part that was predictable from what went before and so was already contained once. Computers could do this trimming. But on the other horn of the dilemma, in every transmission of information accuracy is inevitably degraded, errors creep in—what the engineers call noise increases. (Four decades later, the familiar example of this increase of noise is the inevitable loss in fidelity when one makes a copy of a tape recording or a photograph.) To ensure the content of a message, it must contain enough redundancy so that a part distorted or lost could be reconstructed, at the receiver, from analysis of the rest—filled in, so to speak. Computers could do the filling in, too. And, obviously, the trimming at the transmitter and the filling in at the receiver required exactly the same process—a prediction. Both amounted to treating the message as a series of measurable events in time, and attempting to predict the next passage, the next events (whether a potential redundancy or a potential gap) from what had gone before. "The

---

information is one binary digit, now called a "bit."

The idea of the total amount of information, they then saw, relates to an idea of classical physics, that of entropy. The entropy of a system is the measure of its degree of disorder. That is, entropy increases as the system runs down; entropy tallies the slide toward chaos that is the second law of thermodynamics. The amount of information in a system, by contrast, is a measure of its degree of organization—of order. Information fights chaos, and can sometimes win momentarily and locally even though it must lose eventually and universally. It turned out that the amount of information in a system is simply the reverse of its entropy. And the branch of classical physics called statistical mechanics possessed well-developed mathematical tools for handling entropy.

prediction of the future of a message is done by some sort of operator [that is, mathematical treatment] on its past, whether this operator is realized by a scheme of mathematical computation, or by a mechanical or electrical apparatus," Wiener wrote later.

But this pattern, in turn, was exactly like trying to predict the future position of an airplane from its known past course. Wiener saw that. He did not see the still-further similarity, which leaps out at us, between the computer at the transmitter analyzing the message being sent and the psychological function, in everyday adult speech, that Mead called the generalized other. Yet that was just the kind of unexpected juxtaposition that Wiener loved to make and that gave these ideas remarkable force and plausibility, to laymen and scientists alike.

The essential unity of the problems of communication and control, and of the mathematics for analyzing them, provided intellectual stiffening. About 1947, Wiener and his colleagues, noting that there was not even a single name that embraced the field of control and communication theory, whether in machine or animal, coined for it the term "cybernetics," from the Greek word meaning "steersman." Watt's and Maxwell's "governor" was itself derived from the Latin corruption of that same Greek word for steersman. But parts of the field that Wiener thus branded as his own were by then growing rapidly and independently—particularly information theory, springing from the work of Claude Shannon and others at Bell Telephone Laboratories, and computer theory, in which, among many others, John von Neumann, at the Institute for Advanced Study, in Princeton, had a dominant role. Wiener's discoveries were swept up into the flood of development of what is still, sometimes, called cybernetics, but more usually simply information theory. Wiener was nonetheless the first to see clearly the unity of these themes of control, communication, the message, and feedback. In 1948, his book *Cybernetics: Or Control and Communication in the Animal and the Machine* proclaimed this unity with splendid dash and enthusiasm; it was the most exciting book of the year.

The patterns of feedback appear in unexpected and entertaining places. Biologists soon observed that entire ecological systems are self-regulating by feedback. Relationships between predator and prey are of great interest to conservationists, for example. In parts of Canada, the population of lynxes fluctuates widely over a period of years. Lynxes eat rabbits. When the population of rabbits is low, lynxes starve and their population drops. As the lynx population declines, the pressure on the rabbits eases; they multiply and survive to multiply again; after a while the rabbit population is high enough to sustain many more lynxes than are there. More gradually than the rabbits, the lynxes begin to increase. As the lynx population grows, the rabbit population reaches a peak—but then begins to fall as the lynxes catch up. As the lynx population reaches *its* crest, the rabbit population is already dropping rapidly. Mass starvation leads to a population crash among lynxes—and the cycle is set to repeat.

Similar cycles demonstrate the pathologies of feedback, too. An amusing case of overshooting feedback was recounted by Dennis Flanagan, the editor of *Scientific American*, in a recent conversation. He and his wife went to sleep, one cold night, under an electric blanket. He likes to sleep cooler than she does, so their blanket has two controls, hers set higher than his. But on this night, without their noticing, the blanket has been turned over. After a short time, feeling chilly, she turns up her control. Soon, he wakes drowsily because he's unaccountably warm, turns down his control. In a moment, she, feeling colder than before, again turns up her control. He, beginning to sweat, turns his control down some more. She, shivering, turns hers up full. He, wringing wet, thrashes, sticks his legs out, turns his control off. By now, sleepless and uncomfortable, they start to quarrel.

Flanagan's electric blanket, hunting, resembles what happens with painful consequences in the prices of agricultural commodities, like grain and particularly beef. When wheat, or beef, is low in price relative to farmers' costs, they cut back planned production. Wheat acreage is switched to other crops. Beef herds are allowed to decline. But these adjustments take time—a year at least for grain production, two years for beef. Smaller production begins to drive prices up. Higher prices one year encourage farmers to start building up production again for the next. Grain prices, when oscillating this way, can be driven to sudden excessive swings by such things as the Russian crop failures of 1972 and 1973; then sudden large-scale grain purchases by the Soviet Union, including 400 million bushels in the twelve months to July 1, 1973, drove the price of wheat from $1.76 a bushel in 1972 to a high of $3.95 a bushel in 1973—which in turn drove American production of wheat from a harvest of 1,705 million bushels in 1973 to 2,135 million bushels in 1975, easing the price to $3.55 a bushel that year.

Wiener himself proposed that in animal physiology hormones perform as governors. Messages over the nervous system usually are specific: you have stubbed your left big toe; you reach out your right hand for the pencil. When the message is not specific but is to affect the entire body, hormones rather than nervous signals may be more efficient. In a vivid illustration, Wiener observed that the ordinary communication system in a large mine may be by telephone, from a central office, with all the necessary wiring and equipment. "When we want to empty a mine in a hurry, we do not trust to this, but break a tube of a mercaptan in the air intake." Mercaptans are chemicals that stink powerfully of rotten eggs or skunk. An obvious chemical messenger in the body is adrenaline, produced in a tiny gland atop each kidney, which in sudden rage or fear floods the bloodstream and thus the entire body, preparing it for fight or flight in the most diverse ways—raising the heartbeat and blood pressure, releasing sugar into the blood from the liver, increasing muscle tone and responsiveness, constricting the blood vessels near the skin so that injuries bleed less badly, and so on. A hormonal process with a feedback loop is the insulin system. When the level of sugar in the blood rises, insulin is released, which drives some of the sugar into storage, and as the blood-sugar level falls the level of insulin subsides as well. As a governor, the insulin system can overcompensate. In the condition called hypoglycemia, the body reacts to a slight elevation of blood sugar by releasing so much insulin that the blood-sugar level falls catastrophically and the person goes into insulin shock. The individual suffering from hypoglycemia, if foolish enough to eat a chocolate bar on an empty stomach, may collapse in unconsciousness twenty minutes later.

The sex hormones make up an elaborate interconnected network of feedback controls, releasing or inhibiting the production of chemicals that have widespread and diverse effects on the physiology. The fundamental feedback process here is the interaction between the pituitary gland, situated in the center of the head immediately beneath the brain, and the gonads—the testes or the ovaries. A hormone from the pituitary stimulates the production of sperm in the male, or in the female the maturation of an egg in its follicle in the ovary. The active gonad releases sex hormones—testosterone in the male, estrogen in the female—into the bloodstream. These have diverse effects throughout the body—which include depressing the output of the gonadotropic (that is, the stimulating) hormone of the pituitary.

The hormonal controls in women, built on that loop, are especially intricate and precise. Each month, as the estrogen level slowly builds, the uterus is stimulated to prepare for pregnancy. When the egg bursts from its follicle, a lot of estrogen is released, which depresses the pituitary sharply and thus inhibits the maturation of more eggs (which is why estrogen is used in birth-control pills). If pregnancy does not occur, the controls that keep the uterus prepared must be turned down until the next cycle; but if the egg is fertilized, then the hormonal controls must work the other way, maintaining the uterus and continuing to prevent maturation of more eggs. After the child is born, the breasts of the nursing mother produce still another hormone, which hastens the return of the uterus to normal at the same time that it continues the inhibition of maturation of eggs. Thus, in many primitive societies, the hormones produced by a nursing mother, in combination with the high demand for calories that lactation imposes on her, act as a natural birth-control mechanism, spacing children. In Westernized societies, however, women less often breast-feed (though the practice, health-giving for both mother and baby, is coming back into fashion), and when they do their calorie intake is likely to be high enough so that their bodies can override the control. This description only hints at the complications—all regulated by hormones from egg, follicle, uterus, placenta, breasts, and pituitary balanced against one another in multiple finely tuned feedback.

In the early 1970s, just when physiologists thought they had worked these interactions out, it began to appear that the entire self-regulating network was in part under direct control of the brain. Two groups of molecular biologists in bitter competition, one led by Andrew Schally at Baylor University, in Texas, the other led by Roger Guillemin, first at Baylor and then at the Salk Institute, in La Jolla, California, discovered that the hypothalamus—a primitive part of the brain, lying low in the skull and a fraction of an inch from the pituitary gland—makes several hormones. The pituitary is the target of these hormones. One rules the release of the pituitary hormone that in turn regulates the sex-hormone system, another rules the release of the hormone, also produced in the pituitary, that is a principal element in the growth-hormone system, and so on. The hypothalamic hormones are surprisingly small molecules—a tenth or less the size of the insulin molecule—and chemically simple. They are produced in such vanishingly small quantities that in order to isolate one ten-thousandth of an ounce of one of them in pure form, Guillemin required the brains of a quarter of a million sheep, which he and a crew collected—so fresh they were still warm—on the killing floors of slaughterhouses in five states. At signals from the nervous system which are still not elucidated, a few molecules of one or another of these hypothalamic hormones are emitted to travel the short distance to their targets, their molecular receptors in the pituitary. The hypothalamic hormones are called releasing factors. Unlike the hormones they control—for example, those in the sex-hormone system—the releasing factors affect nothing but their specific targets. They have no known side effects. Yet they are the master keys to several of the most powerful systems of the body. Their discovery brought Guillemin and Schally shares in the Nobel Prize in physiology or medicine for 1977.

Wiener, Shannon, and von Neumann were intent upon the similarities, practical and theoretical, between the brain and computers. In the middle and late 1950s, however, one of the most distinctive and important patterns of information and feedback was puzzled out in the biochemical machinery of the cell. The research was done by molecular biologists, chiefly in Cambridge, England, in Paris, and in Boston.

The question—in the literal sense, the *vital*

question—was how the hereditary information, the information carried on the genes, dictates the making of the organism. Genes are made of DNA. The cell also contains a stupefying variety of other molecules, comprising fats, sugars, and proteins, but it was clear that what the genes do is act as blueprint to specify how proteins are assembled. The proteins make up the working machinery of the cell and thus take care of everything else except the blueprint. But DNA and proteins are chemically altogether dissimilar. By 1954, when the structure of DNA molecules had been worked out, that vital question could be restated: How does the information in the strand of DNA specify a strand of protein? More particularly, the strand of DNA is made up with four kinds of chemical subunits, called bases, while the strand of protein is made up from twenty kinds of chemical subunits, called amino acids. How did a message written in an alphabet using four letters, the four DNA bases, get read off and translated into a language expressed in twenty characters, the amino acids?

By 1961, after years of bafflement, the outline of the answer emerged. The information on the strand of DNA is first transcribed onto a strand of a closely similar substance, RNA. This strand of RNA is called a messenger. It carries the information to the places in the cell where protein molecules are synthesized, and there the base sequence, now in the messenger RNA, directs the hooking together of specific amino acids to form the protein strand. Three bases in a row are needed to specify a single amino acid. Two in a row (any of the four kinds in the first position, any of the four in the second position) could not specify more than sixteen of the twenty amino acids. Three in a row give $4 \times 4 \times 4 = 64$ possible words—they're called "codons"—in the four-letter base language. In fact, sixty-one of the possible codons specify individual amino acids; there are many synonyms. The other three codons are punctuation marks. So much for the transcription and translation of the genetic message.

The cell must also regulate the expression of the genes—must turn them on and off, or else the cell would be making every kind of protein it can make, all at once and all the time. The regulation of protein synthesis uses feedback loops at several levels. For example, a particular species of bacteria may be able to use any one of several different sugars as its food. To do so, it must chop up molecules of the sugar into smaller pieces, releasing energy in the process and getting the pieces to use as building blocks in making other substances. The digestion is done by an enzyme, which is a protein molecule, specified by a gene—a different model of enzyme, and so a different gene, for digesting each kind of sugar. When one kind of sugar is not present, the bacteria don't make the enzyme to digest it. It turned out that a special protein molecule, called a repressor, attaches to the DNA itself at the beginning of the gene for that enzyme, blocking transcription of messenger RNA. When the sugar *is* available, a molecule of it plugs into the repressor like a key into a lock. The repressor changes shape and falls off the DNA. Transcription of the gene into messenger begins. As soon as the message is transcribed, its sequence begins dictating assembly of the sequence of amino acids. Within three minutes, the first molecules of the enzyme are made and are beginning to chop up the sugar. The bacteria grow and divide, repeatedly. When all of that sugar is consumed, the repressor molecules are no longer deactivated; they reattach to the DNA and the bacteria waste no materials and energy making any more of that enzyme.

The repressor model is just one of many different control systems found in cells. The details vary widely, and in higher organisms are bafflingly complex, but nonetheless feedback and the processing of sequence information are essential to the life of every cell of every organism.

The patterns and terminology of feedback seem all but indispensable in describing these controls. Ironically, though, the scientists who first teased out the intricate interrelationships did so almost without reference to cybernetics and information theory. Only when the most fundamental discoveries had been put together into the pattern was it recognized for what it is.

Today, more than fifty years after George Herbert Mead pointed to the patterns of feedback in the learning and use of language, social psychologists are taking the obvious next step. Using videotape and sometimes the technology of espionage—tiny, pin-on microphones with built-in short-range transmitters, received at voice-actuated tape recorders—and investing, as well, the patient labor to transcribe and study the tapes, psychologists have programs under way that record every interaction in the first year of an infant's preverbal and verbal development.

In ten years of deeply thoughtful, deeply responsive studies, Selma Fraiberg, who is a psychoanalyst at the University of Michigan and specializes in children, compared the development of blind and sighted infants in their first months. We normally read a baby's emotions mostly from its face—smiling or crying, of course, but also, long before speech, suggesting shades of alertness or boredom, playfulness, puzzlement, searching, wanting. The baby's expressions that can be read as feelings form crucial feedback loops in the second-by-second interactions with its parents. One all-important instance: from the first weeks, the baby's looking the mother in the eye is a potent sign and trigger for the mother's loving response to the baby. Eyes catching eyes is the preliminary even to smiling. But babies born blind show much less facial expression than do sighted infants. By about nine weeks old, the normal baby smiles at the sight of a face—and gets a smile back. Blind babies even need to be taught to smile, by nuzzling and tickling games. Their faces are disconcertingly blank. Of course, they never look the adult in the eye, and even to the mother this may seem cold and unfriendly. Fondling a favorite toy, they have their heads turned away in a posture that makes them seem bored.

Fraiberg noticed that in the absence of the child's facial response, adult observers and even parents show much less facial animation themselves. She found—first by becoming aware of her own reactions, and much to her chagrin—that adults *talk less* to blind infants. They're failing to get the feedback from facial expressions that keeps the playful chatter going. Yet of course the blind baby feels shades of emotion and desire as does the sighted—and expresses them. "But we have to turn our eyes away from the face to discover them," Fraiberg wrote in 1974. "To do this is so alien to normal human discourse that we might not have discovered the other signs if we had not been looking for something else." She and her associates were trying, in fact, to follow how blind babies learn to coordinate their movements and to grasp objects—and how they learn to understand that an object removed from their touch continues to exist and may be recovered.

Analyzing babies' use of their hands, with the aid of thousands of feet of film, which they projected at one-third normal speed, Fraiberg and her colleagues began to see that the emotions were expressed in the hands. As they became sensitive to expressions not in the face but in the hands, they "began to read them as signs and respond to them as signs." (They also began to teach parents of blind babies to read the signs in the hands.) A telling example for the later development of language by sighted and blind alike: when blind Robbie was eight months old, they brought his musical dog within easy reach of his hands and played its familiar music. Fraiberg wrote:

> He will not reach for the toy. Does he want it? His face does not register yearning or wanting. But now as the music plays we see his hands in an

> anticipatory posture of holding, grasping and ungrasping.
>
> At nine months of age we ring our test bell within easy reach of Robbie's hands. The bell is a favorite toy for Robbie. He does not yet reach for it. Does he want it? We watch his hands, and then we see the hands execute a pantomime of bell ringing as he hears the bell "out there."

With a neatness and complex symmetry that one grows to expect in the patterns of feedback, the forty years of the rise and flowering of this theme and its related concepts have been the years when, also, the enterprise of science has been subjected to scrutiny, analysis, and close and intense criticism. Unlike any other approaches to acquiring knowledge, science in the doing is a self-regulating system. Recall Sir Peter Medawar's description: "Scientific reasoning," he said, "is a constant interplay or interaction between hypotheses and the logical expectations they give rise to: there is a restless to-and-fro motion of thought, the formulation and reformulation of hypotheses, until we arrive at a hypothesis which, to the best of our prevailing knowledge, will satisfactorily meet the case. Scientific reasoning is a kind of dialogue between the possible and the actual, between what might be and what is in fact the case." One narrows the gap. One picks the pencil up. Feedback is a model of science itself.

6

Modeling

On December 12, 1966, as a promotion stunt, *Scientific American* ran a big ad in *The New York Times* to launch the First International Paper Airplane Competition. The ad, mock-serious, compared the design of supersonic transports to the creation of paper planes, invoked Leonardo da Vinci as the patron saint of paper planes, and offered prizes (trophies called The Leonardo) for winners in four categories—duration aloft, distance flown, aerobatics, and origami. The contest took off. Within hours, the magazine's offices were crowded with journalists. Three days later, the San Francisco *Chronicle* gave the contest a banner front-page headline; within a fortnight, editorials applauding the contest appeared in more than a hundred American newspapers. By the day of the Final Flyoffs, at the New York Hall of Science on February 21, 1967, the entries totaled 11,851 paper planes, from twenty-eight countries including Liberia and Switzerland. Makers of the planes included schoolchildren, scientists, and aeronautical engineers. The smallest plane measured .08 by .00003 inches while the largest was 11 feet long. Many of those in between were elaborate, bizarre, inventive flights of fancy.

The event, with the surprising and enthusiastic response to it, captures on the wing a revealing paradox about how scientists and technologists think. The first limb of the paradox: model-making is a profound and instinctual human response to comprehending the world. Everybody has made paper airplanes, sure. But everybody makes models of many kinds, and all the time. Children make models of the physical and technological world they know, endlessly building and rebuilding with clay, mud and water, blocks, bricks. Children model the *social* world around them and teach themselves how to get along in it, experimenting and exploring without risk, by their play with dolls. This process is so crucial to children's development that one of the most important tools of child psychiatry, when something seems to be going wrong with the way a small child handles the social world, is the close observation of the model of his life the child offers in playing with a family of dolls. Adolescents will spend countless patient hours constructing models, historical or functional, of wood and paper and string—and this just at the time of life when a scientific vocation usually takes hold. Throughout adult life, too, people make models, though often in ways that may not be immediately obvious. Scientists and technologists are unusual, here, only because they make models with formal and deliberate seriousness, as professional tools.

The second limb of the paradox: modeling, however serious, enshrines an element of play. Watching a child, one turns that observation the other way around: for the child, absorbed in play, modeling has an essential aspect of seriousness—it's a way of grasping the way things are. For the scientist or engineer, conversely, the seriousness of modeling retains something of the youthful delight. Scientists are incessantly saying to each other "let's play around with that"—and modeling is the quintessential way of playing with the way things might work and might be. The play and the seriousness—their fusion generates the energy that *Scientific American*'s paper airplane competition tapped. Of course, the competition drew enthusiastic response. The contest, if one thinks about it, invited participants to have fun modeling *modeling.*

Nicholas II, Czar of all the Russias, gave his wife, every Easter, a decorative egg. The Russians have to this day a pleasant Eastertide tradition of giving each other painted wooden eggs—simple peasant crafts. The emperor's family had *their* Easter eggs made by the court jewelers, Fabergé—fabulous works of the miniaturist's art. For Easter of 1901, Nicholas gave his wife an egg made of gold, silver, gems, and enamel, 10¾ inches tall, that opened on a golden hinge. Inside nested a model of an express train from the Trans-

Siberian Railway—locomotive, tender, five passenger cars including a traveling church, to be taken out and run under its own clockwork power along the table. The bodies were of platinum and gold, the engineering finished in every detail: doors that opened, windows sliced from rock crystal, the interiors of the cars completely furnished down to the No Smoking sign, on the wall of one, with lettering so tiny it could be read only with the aid of a microscope.

Listen to scientists talk about their work. Whatever the field—theoretical physics, chemical engineering, ecology, social psychology, economics—they talk about their models. They use "modeling" in many overlapping senses. But the one thing they never mean is something complete in every possible detail like Czar Nicholas's toy train. In science and engineering, however else one may speak of modeling, what's always meant is that the model leaves out the trivial details to embody the important, the useful, the crucial features of the thing modeled—the features that make a difference. Listen to Nobel Prize–winning scientists talk about *their* work. As part of the Nobel ceremonies in Stockholm in December each year, each winner gives a lecture on his work—and a surprising number of these lectures have dealt with modeling. "The art of model-building is the exclusion of real but irrelevant parts of the problem," said Philip Anderson, a solid-state physicist at Bell Telephone Laboratories, in his Nobel lecture in 1977. Anderson's explanation was also a warning: "Model-building entails hazards for the builder and the reader," he said. "The builder may leave out something genuinely relevant; the reader, armed with . . . too accurate a computation, may take literally a schematized model whose main aim is to be a demonstration of possibility."

In Anderson's physics, a model is likely to be expressed as a set of equations, a scant pageful of symbolized relationships. At the other extreme, we were told by Ralph Nelson, a physiologist at the Mayo Clinic, in Minnesota, "I thought the hibernating bear was a perfect model for studying the problem of how human beings metabolize fats and proteins—because the bear hibernates without his temperature dropping, and can starve for five months without burning his protein, which humans cannot do for five days." From a set of equations to a living animal: yet both can be models of something else, in the sense that scientists and engineers speak of models, because each in its way isolates the features of the something modeled which make a difference. In the terms of chapter 3, "Change," what one models in science and engineering are *parameters.* Model-building, therefore, can be one way to perform an experiment. Curiously, model-building can also be a way to create a theory.

The performance of objects. The behavior of elaborate systems. A kind of theory-making. Modeling performs at these three levels.

Will it fly? The paper airplane is the imaginative prototype of the model that's used to predict the performance of an object. The first wind tunnel was built in 1871, by two British engineers, who used it for experiments with wing shapes. The Wright Brothers, in the years before their first powered flight, built gliders in their bicycle shop in Dayton, Ohio, and experimented with them in the steady ocean winds along the dunes of Kitty Hawk, North Carolina. The glider the Wrights built in 1901 turned out to be unstable and difficult to control. Rebuilding it taught them that the available data on the aerodynamic performance of kites and wing surfaces were dangerously wrong. (The man who had compiled the tables had died in a crash of his hang glider.) Back in Dayton, the Wrights constructed their own small wind tunnel; with linked balances made out of bicycle spokes, they measured the forces that the airstream exerted on more than sixty model wings of different shapes and thicknesses. From their own data, the Wrights designed a new glider for 1902. It had a 32-foot wingspan, half again as long as their glider the year before, and weighed over 100 pounds. It was the largest thing that had ever flown. The Wrights also devised vertical control fins, the first ever tried. That season at Kitty Hawk, the Wrights broke all records for duration aloft. The success of the design, and the skill they had acquired at flying a glider in dynamic balance—like a coasting bicycle—led, a year later, to the first powered flight.

The Wrights were more than inspired bicycle mechanics. The performance of a wing surface depends on four parameters—weight, thrust, lift, and drag. A wind tunnel allows two of these to be measured, lift and drag. Lift and drag both depend on the shape of the wing—its surface area, top and bottom, its curvature, its thickness, the relation of its length to its width—and then upon its angle of attack against the airstream. A present-day engineer would measure the lift and drag on a model wing directly, with gauges sensitive to the small variations in the forces operating on the model; he would then scale up. The Wrights measured lift and drag indirectly. Their bicycle-spoke linkages enabled them to ascertain the forces in relation to each other and to a flat surface in the wind tunnel, to be compared with measurements they had made on a full-sized flat surface in the steady wind at Kitty Hawk. The Wrights' aerodynamic innovations have recently been reexamined by an American engineer and inventor, Frederick Hooven. (Hooven happens to have been the winner in the duration event in the First International Paper Airplane Competition.) Hooven concluded that the Wrights, by the intuitive good sense of their measuring devices, "avoided the worst pitfalls of model testing on a small scale: the difficulty of measuring small and probably unsteady forces and the effects of scale that result from the viscosity of the air (which the Wrights probably were not even aware of)."

The advantages of modeling an object's performance are obvious: models are a means to trial and error—a way to make mistakes safely and relatively cheaply. The tank tests of a one-eighth scale model of the hull of a new boat for the America's Cup competition—a model less than 6 feet long at the water line—may cost $80,000, but the completed yacht will cost a million before her owner buys her first suit of sails. Then there are devices that must work right the first time, and can only be tested by some form of model. At the Jet Propulsion Laboratory, in Pasadena, California, James Blyn, who is both an aerospace engineer and a specialist in computers, put models of the Voyager—the unmanned spacecraft designed to fly past Jupiter and Saturn—into the lab's computer. The computer took in the model first as a string of numbers; Blyn then programmed an animated display by which the computer put a drawing of the Voyager onto the screen so that it could be turned on command, examined from any angle, its behavior simulated, changes tried. "The modeling that goes on with the computer has to do with objects in space—things that are real, or simulated as potentially real," Blyn said recently. "This gives you a picture of what they would actually look like—much more readily interpretable than a list of numbers, and a lot more interesting. Animation makes the program interactive with the engineer; it's much more alive, in that you can put changes into your model much quicker and see the results of the changes come up on the screen much quicker." Blyn's computer program then flew the model Voyager through a model solar system past a model Jupiter. "We picture the image that the Voyager is going to see looking at Jupiter and the moons of Jupiter as we go past. And one thing that that enabled us to do was to catch some of the problems we were going to have with the camera—like parts of the spacecraft itself getting in the way of the camera at certain angles—to catch these things quickly, and adjust the commands for positioning the camera that we were going to send up."

Modeling's problems are less obvious than its uses—until one recalls the scale effect. That competition entry of a paper airplane that measured .08 by .00003 inches was a joke, yes, but the joke's point was that the model was a sophisticated try for the prize in the duration event—scaling the plane down toward the surface-to-weight ratio of a feathered seed or a dust mote. (It didn't win.) Weight, thrust, lift, drag—but the two of these parameters that are measured in wind tunnels both relate exclusively to the surface of the model. Yet engineers found early that results with models in wind tunnels did not match what happened with full-sized aircraft.

The reason was a different sort of scale effect. In 1920, Max Munk, at Langley Research Center, in Virginia, proposed that since the model in the wind tunnel is reduced from the real thing, the characteristics of the airflow around and past it will be different unless a compensating adjustment is made in the air itself. If a model were, say, a twentieth of full size, then it should be tested in a flow of air of twenty times normal density. Munk was invoking observations that had been made by Osborne Reynolds, an English engineer, in 1883. Reynolds had determined by experiments with liquids in pipes that the smoothness or turbulence of the flow of water—for example, past an obstacle in a stream—depends on several factors. Turbulence increases with the velocity of the flow: a rock in the stream has more eddies around it when the water is moving faster. Turbulence also increases with the size of the object intruding in the flow: the rock disturbs the stream more than a pebble does. Turbulence increases with the density of the material in the stream: a "thinner" fluid is not so perturbed by the rock. But turbulence decreases with the viscosity—the tendency to stick-to-itself—of the fluid: honey flows around the rock more smoothly than water.* And smoothness or turbulence of the flow of air crucially affects the performance of a wing. Thus, Munk said, the smallness of the model entailed a reduction in turbulence, which could be restored by building a closed tunnel where the air was recycled under pressure, to increase the density of the airstream. The first variable-density wind tunnel was built in Langley in 1923, by the National Advisory Committee for Aeronautics—which became the National Aeronautics and Space Administration in 1958. Munk was proved right.

To model supersonic flight requires heroic measures. In some tests, the air is cooled with liquid nitrogen to −300° Fahrenheit, which reduces the viscosity—the fourth of Reynolds's factors—so greatly that the turbulence corresponding to supersonic velocities can be reached at lower speed and so at a relatively low cost in energy to drive the airstream. Similar problems of scale bedevil the tank testing of model ships. To scale the drag of the liquid to the surface of the model hull, strips of sandpaper are glued to the bow, or pins are studded into the model, below the waterline. Architects, in order to design room layouts that will allow efficient ventilation, model the movement of air through miniature rooms not

---

*Turbulence, as a relationship among these four factors, is often expressed by multiplying the first three—the Size of the obstacle, the Velocity of the fluid, and the fluid's Density—

$$S \cdot V \cdot D$$

and then dividing this product by the fourth factor, the viscosity of the fluid—the factor with which turbulence varies not directly but inversely. The result is a measure of Turbulence:

$$T = \frac{S \cdot V \cdot D}{v}$$

This result is called a Reynolds number. The useful fact is that the characteristics of flow in any two systems, of whatever size and composition, will be similar if their Reynolds numbers are similar. Getting wind-tunnel results that work is a matter of adjusting the factors so that the Reynolds number matches that of the full-sized situation.

with air but on a flow table in a current of water. Engineers modeling a new dam contend with the same problems of the flow of liquids—but here, Reynolds is compounded by Galileo, for a dam, like a cathedral or a skyscraper, also presents the more familiar danger of the mass of its materials increasing with the volume, by the cube, as the model is scaled up to the real structure. The engineer may load the model of the dam with sandbags; he may simulate the pressure of the deep lake behind the dam by replacing water with mercury.

A map is a kind of model. The child's idea of a map is that the best is the one drawn to the largest scale—the map that shows the most, not just the road he lives on, but his very house. Push that idea to its absurd conclusion, though, and the map becomes as big and as detailed as the territory itself—but long before that, flapping in the breeze, of course it loses its utility. Yet in other contexts, the mistake of running the map into the ground can be easy to make. "Ah, ha!" says the sales manager. "With our new computer I can at last take a look at the performance of each salesman with each item in our list each week of last month"—and the computer obligingly delivers a printout sixty inches thick.

"The exclusion of real but irrelevant parts of the problem"—perhaps the most elegant instances, generally known, of maps that obey Philip Anderson's prescription for model-building are those of the New York City subway system and the London underground. The cartographers' deliberate distortions of the actual geography—extreme simplifications of the real spacings of the stations and windings of the tunnels—produce in each of these maps a handsome pattern excluding all but the relationships and directions of the stops, the lines, and their intersections. In the process, Manhattan is broadened to the shape of a Ping-Pong paddle, the Thames's meander stylized, information useful to motorists or pedestrians rigorously ruled out: in the result, each is a map not of a topography but of a network with knots in it—a topology. And the maps of the ancients, with monsters and fabulous animals at the bounds of what the cartographer knew—they were models too, perhaps, but laid out to a rule the converse of Anderson's; that is, by the inclusion of unreal but psychologically relevant parts of the problem.

A model of the behavior of an elaborate system—the second level at which modeling is essential to science and technology—begins as a map. Complex data, whose interactions are not fully understood, are selected and put into relation—more or less tentative—to each other. Then the map, so to speak, is turned on and set into motion: the changes in certain parameters are introduced into the system and their interactions followed. Where the model of an object attempts to predict performance, the model of a complex system tries to predict *outcomes.*

To the southwest of the Japanese Archipelago is a fissure 300 miles long, which splits the main island of Honshu from the smaller Kyushu and Shikoku. This fissure is the Inland Sea of Japan. Its outlets are three straits. The three islands that bound it are mountainous, drained by many rivers, their coastlines deeply irregular. Japanese history began around the Inland Sea. Fourteen million people now live on its rim. Scattered across it are three thousand islands and islets, some dividing the sea into smaller seas and partly enclosed bays. The sea is shallow. Until well past the Second World War, the waters teemed with fish. Now, long stretches of the shore are thronged with industry—steel-making, shipbuilding, petrochemicals—while other parts are preserved as national park, forever wild. By 1971, pollution of the Inland Sea was recognized as a grave problem. A research institute was set up. Scientists began building a hydraulic model of the Inland Sea, to a scale of 1 to 2,000—a reproduction that measures

8,970 square yards and fills an airplane hangar. Water can be pumped through the model to simulate the complex interplay of the tides. One day's tidal action takes nine minutes on the model. Streams of dye or small floating colored discs are released to simulate the drainage of pollutants from the shore. The model is observed from catwalks high above it, and each variation—each experiment—photographed. Interrelationships emerge. Remedial and preventive steps are invented, tested, their practicality assessed, their effectiveness related to their cost.

The two most complex systems for which models are now being built are the weather and the economy. The difficulties of predicting either with accuracy are notorious, and the reasons for the difficulties similar. It's simple, if laborious, to accumulate and map vast amounts of data about what's happening at many particular points in either system—statistics from weather stations in Greenland, Bermuda, Key West, and so on, or of interest rates, consumer optimism, machine-tool purchases, and so on. By taking the measurements repeatedly—perhaps hourly for the weather, quarterly for the economy—the rate of change at each point on the map can be charted. But to determine how the changes going on at all the different points on the map interact with one another is obviously complicated. To go further, to see how the *rates* of change are themselves changing, and how these changes interact so that the performance of the entire system slows, stops, turns in the other direction, is immensely more difficult still. Yet nothing less sophisticated will be much use at all.

Models are therefore indispensable. To predict the behavior of a hurricane along the east coast of North America could only begin to be possible with the day in 1760 that Benjamin Franklin realized, from an exchange of letters between Philadelphia and a friend in Boston about the passage of a recent destructive storm, that although the winds of the hurricane had struck each city from the northeast, the entire storm had been moving in the opposite direction, from southwest to northeast. Franklin had the beginning of the model which, since the Second World War, has made it possible at least to anticipate and avoid disasters like the hurricane early in September 1936 that killed more than two hundred people from the Carolinas to Iceland and Newfoundland. Similarly, economists had to have at least a rudimentary model of the world economic system to predict the degree of inflation that resulted from the quadrupling of oil prices by the Organization of Petroleum Exporting Countries at the end of 1973 and the series of further hikes since. To progress beyond the warning of gross and imminent disasters, whether of the weather or of the economy, has required models of increasing intricacy and size. They have begun to be possible only because of computers.

Every college student who has gone through a phase of playing Monopoly obsessively for a few weeks has hankered for a version of an economic game that would resemble the real world. When I was at university, after the first year Monopoly became so insipid that some friends and I devised a new board game which introduced a stock market, allowed formation of corporations that also appeared on the board as players, and so on. We played it once, but abandoned the attempt when no clear winner was in view after fourteen hours. Moves took too long, negotiations between moves took still longer, we could not preserve secrecy, partial ignorance, and the unexpected. We needed a computer to play the game on. Now, in the late 1970s, at least a dozen competing models of the American economy are established on computers. A model is being built of the world economy. These models started in the fifties from just the kind of play with computers that my friends and I speculated about and that young economists were beginning to explore; indeed, an influential book at the time was *Theory of Games and Economic Behavior,* by John von Neumann, who was a profoundly original intelligence in information theory and the use of computers, and Oskar Morgenstern, an economist. Present economic models chew up incredible masses of data; by multiple, repeated comparisons of trends they establish which of the data merely reflect more fundamental changes, and which appear to penetrate most closely to the controlling parameters. The models then allow the predictions to be compared with what actually ensues—in order to correct and refine the model. Such models are called econometric. Their development is without doubt the most important change in the way humankind runs its affairs that we will see this century.

A model that will allow sensible and humane efforts to control and adjust the performance of the national and perhaps the world economy is highly desirable. A computer that turned the same sorts of data to the increased control of individuals would be pernicious. Proposals have repeatedly been made for the building of a computer copy of the economy that will include every recorded transaction—every check through every bank account, every bill charged, every stock or bond traded—not only to base forecasting on more complete data but to monitor fraud, eliminate tax evasion, and so on. Yet the temptation to run the economic map into the ground may prove largely self-defeating. For one thing, what's called "the alternative economy"—services performed for cash, second jobs paid in cash, barter clubs where doctors, lawyers, housepainters trade services with other members not for money but for redeemable barter points—has been growing vigorously in the past decade; it would multiply manyfold as official controls got tighter. More fundamentally, a duplication of the entire economy's transactions on a computer would not of itself be a model. The parameters—and so the usefulness—would be capsized in their mere consequences, drowned, just as Philip Anderson warned, in the "real but irrelevant."

Where a model of an object permits inexpensive trials and safe errors, a model of a complex system may become an exercise in theory-making. Indeed, that incessant interplay between prediction and correction of the model is the chief characteristic of all theorizing in science. Yet some models are theories in the full sense and nonetheless are far simpler than the world's weather or the world's trade on a computer.

In the spring of 1948, Linus Pauling made a model that embodied the results of work and thought he had been pursuing for more than ten years. His model was a theory about an important aspect of the nature of life itself, yet was as naïve as a paper airplane. He was in Oxford at the time, giving a series of lectures on the bonds that hold atoms together to form molecules of one chemical substance or another. He caught cold and went to bed. Pauling was the world authority on the nature of the chemical bond and the physical structures of molecules. For more than a decade, he had been puzzled about the structures of protein molecules—among the largest, most complex, and most important of the molecules of living organisms, including as they do the enzymes, the antibodies, many of the hormones, and the oxygen carriers like hemoglobin. By 1948 it was well known that any protein molecule is a long chain whose links are some sequence of component molecules, the twenty-odd amino acids, which are similar but not identical. Pauling's laboratory, at the California Institute of Technology, had accumulated extremely precise data on the chemical bonds within amino acids and between them. These dimensions are tiny almost beyond our imagining. The length of the bond linking two amino acids, for example, is five-billionths of an inch. Nonetheless, by the technique called X-ray crystallography, the spacings of the repetitive layers of atoms within a crystal of a substance can be projected, in large, as a pattern of spots on a sheet of film. The spacings and intensities of the spots can be read back into the crystal: with the pattern and some heavy mathematics, the three-dimensional placement of each atom in the structure can be determined. In this way, Pauling and his colleagues had gotten the exact spacing of amino acids and the lengths and angles of the bonds between them along the backbone of the protein chain.

"In Oxford, in bed with a cold, I got bored with reading detective stories," he said in a conversation in the fall of 1978, and went on, "I thought, Why don't I think about the structure of proteins? So I took a sheet of paper and constructed, carefully, a drawing of a polypeptide chain, the bond lengths and angles correctly shown." As he told the story, he uncapped his pen and drew a zigzag sketch on a piece of plain paper, to show roughly how he had gone about it. The sketch was the backbone of the protein chain. Each amino-acid component stretched three atoms along the backbone—zig-zag-zig, next one zag-zig-zag. At each bond where they joined, as Pauling had known in 1948, the atoms on either side had to lie all in the same plane. That meant that the chain could turn corners only at each third atom, a carbon atom called the alpha carbon, in the backbone. As we talked, he finished the sketch, and picked up and creased it through one of the alpha carbons—just as he had done in Oxford, thirty years before. "I folded the paper through the alpha carbon atom, here. I repeated that fold, in parallel to the first one, several times, through other alpha carbons." He had been trying to find a way to coil the chain, turning at the alpha carbons, so that bonds would be formed between atoms several times before he got the coiling right. "I finally found how to fold it so that the hydrogen atom on the nitrogen, here, just points straight towards the oxygen atom, sticking off the other carbon"—not the alpha carbon—"at the fourth amino-acid residue removed. And it was at the right distance for the bond to form."

With a colleague at Caltech, Robert Corey, Pauling went on to build three-dimensional atomic models of the alpha helix. They made the models by a system of components they had invented themselves, using bright-colored knobs and balls that represented the atoms and that functioned almost like an analogue computer to prevent mistakes. The shapes and sizes of the knobs were accurate to within a thousandth of a centimeter to represent precisely the sizes at which atoms begin to get in one another's way in different combinations; the joints where they fitted together had the correct bond angles and distances built in. "If you have a model, you know what the permissible structures are," Pauling said to me. "The models themselves permit you to throw out a large number of structures that might otherwise be thought possible. But then, I think that the greatest value of models is their contribution to the process of originating new ideas." Was modeling a form of theory-making? "Yes, I think so," Pauling said. "You remember that professor of mathematics who was asked by a colleague whether a student should be studying mathematics or studying a language. And he replied, 'Mathematics *is* a language.' I would say that models constitute a language, too. They hold information and communicate it. Also, the construction of the model may represent the development of a theory—and with a model, it needs to be a precise theory. If you have fuzzy, vague ideas, then you find that you can't build a model, because a model must be precise."

Late in the winter of 1950–51, Pauling announced the structure he had found in proteins, calling it the alpha helix. The model that began with a zigzag on a sheet of paper, folded into a tube, was one of the discoveries that brought Pauling the Nobel Prize in chemistry in 1954.

A Conversation with Lawrence Klein

Modeling the world's economy is a million-dollar Monopoly game.

The most comprehensive models of the American economy are those on the computer of the Wharton Econometric Forecasting Association at the University of Pennsylvania. The chief architect of these models—and one of the most influential economists in the world—is Lawrence Robert Klein, Benjamin Franklin Professor of Economics at the university. One snowy morning early in 1979, I visited Klein at his office in Philadelphia. The political upheaval in Iran was at a peak, and the shutdown of Iranian oil production and its effect on the economy of the United States and the world were problems worrying everyone. Klein had just returned from a trip to Washington. He was gray-haired, of middling height and open countenance, in his late fifties. He had built his first econometric model of the United States more than thirty-five years before, he said—a matter of a few equations; a scant scattering of statistics about production, employment, and the levels of business activity; and several weeks with a slide rule and a mechanical desk-top calculator. "Models in those days were tiny," he said. "You could put all of it on a page or two. And now, in order to display all of the models you would need a whole book."

What does an econometric model consist of?

"Well, a model is first a data set, and then a set of statistically estimated equations, based on that data set," Klein said. "And then today's model would also include the computer software"—the program of instructions—"that would tell how to apply it, and, in some cases, how to solve it." In other words, how to set up the model and make it run? "Yes," Klein said. "In technical terms, an economic model nowadays is a system of simultaneous finite-difference equations—where 'simultaneous' means that you are concerned with the interactions of things that are going on simultaneously, and 'finite difference' means that the equations, because they are handling statistical data, are a stepwise approximation to complicated sets of dynamic differential equations.

"But we start with a data base—which is the past history of the economy, in very great detail. It is now recorded on magnetic tape or discs, whereas a generation ago it was in a deck of punched cards, and the generation before *that* it was on long accounting sheets." The data included production figures, employment, wage rates, interest rates, prices, taxes at all levels, and so on and on—and these things not just nationwide but region by region and industry by industry or commodity by commodity. The problems of getting accurate data, in fine detail, were sometimes great. "The model we have here for the U.S. economy is a thousand-equation model," Klein said. "And every equation has, maybe, half-a-dozen parameters."

Wasn't he running the map into the ground? A model, I said, was supposed to leave some things out; wasn't he attempting not a model but a simulation of the economy?

"This *is* a simulation. This is a model simulation," Klein said. "You see, my view is that the economy is really definable in fine detail as a giant equation system. For example, you go to a restaurant and buy a meal. What you order depends on your income level—you could buy a more or less expensive meal. You are looking at the menu and thinking of the prices, and your income, your bank account, what else you expect to be doing and eating and spending in the next few days—it's a complicated set of decisions. Our equations are all multivariate. There are very few economic equations—decisions—in which one thing depends on just one other. It usually depends on several others. When the economy is viewed as a giant equation system, all those decisions at every moment all together, it has billions and billions of equations each with many parameters. So we, here, with the model, are making an enormous simplification—but for the human mind it's still a complicated thing to deal with a thousand-equation system."

Then Klein said, "I would put it this way. Any model is an approximation to the universe you're dealing with—and there are many ways of approximating. Some models approximate it with big-industry details. And some are very simple models. And sometimes we model pieces of the economy. Right now, I have the following models in the computer that I'm using in one way or another. A model of New York City; a model of Philadelphia; models of several other cities; models of certain individual states; models of the United States on a fine-time basis, say on a quarterly basis, for immediate predictions; models of the United States on an annual basis, which is the one for predicting long periods into the future. And we have models of other countries, twenty-six or twenty-seven other countries and regions. And we have all those put together into a world scheme. We have models—the most interesting model we have ever built is a model of the Soviet Union." He must have had problems with the data base and the equations there, I said. "Oh, great problems," Klein said. "And we are just getting our feet wet with modeling China. That one's not very far along—but there does exist what you might call, if not a first, a zero-order attempt."

One use of the models was to look back at economic history, Klein said—to learn the lessons of the past. "Reexamining what *might* have happened *if*." And also, I asked, by setting up, say, the 1970 data and then running the model to predict 1972, you could see how accurate your model is? "Yes, that's exactly what we do," Klein said. "We validate models by seeing how they interpret past history—historical trends. *And* we see how they interpreted *singular* events. We have spent a lot of time on, say, Can we model what happened in 1929—can we model the 1929 crash?"

And can you?

"Well, to a certain extent," Klein said. "All models try to sort out what you would call the signal from the noise—to use the language of radio engineering—and the noise component, or what you call the chance, the random or probabilistic component, may sometimes be quite large. And I think there was a big noise component in 1929—besides the real systematic problems of the economy."

In other words, the chance factors in an unstable situation ganged up on the economy?

"Right. It was, maybe, a one-time overextension of capital gain, people playing the stock market, and the capital-gain bubble burst and brought down the whole system."

So the simple-minded popular explanation of 1929 perhaps turns out to be the correct one, after all?

"Right," Klein said. "Now, to take another singular event, I've been making simulations for the whole last two weeks, for Washington agencies and our own group here, on what the consequences are of the Iranian pullback from the world oil market. You might put the question, 'What are the consequences of the Iranian political turmoil?' But we translate that—for this purpose—into a shortfall of so many millions of barrels of oil, and the Iranians reneging on debts, and cutting back on imports from the industrial nations and cutting their military-equipment orders, and so on. And what does it all add up to? A model let us make those relative additions and subtractions."

If a model was to be any use, I said, it must be able to help predict the effects of one-time events as well as general trends.

"That's right," Klein said. "And then we are being asked the question, 'Suppose the Iranian disturbances spread throughout the Middle East. How serious could that be?' "

How fast could he generate answers to such questions?

"In principle, we could do it in seconds," Klein said. "But the models we are using for that purpose are very complicated. They take time to set up. It has to be a world-trade model. So that's a model of twenty-seven countries and areas. And you have to see what's going to be the German reaction, the French reaction, the Japanese reaction, the American reaction, and so forth, and we look at each country individually to adjust their import prices, their oil consumption, and so on. Then we put them all together in our world trading system. So the setup time takes us, maybe—a week. After we set up each piece and test out each piece, then we execute. Push the button. And the execution time might be a hundred seconds or less.

"The Secretary of Defense went to the Middle East to talk to the Saudis and others last weekend, and before he went we had a request for, maybe, four different scenarios, of increasing levels of disturbance—just as part of the background information he was using. We finished three of them before he left. The fourth one had to be cabled to him there—saying what if movements like the one in Iran spread to other Middle Eastern oil producers and there was political turmoil and cut-off by those."

And could he say what the answer was?

"You see, we say that if it's confined to Iran, at this stage, then the result should not have the dimension of a world recession; but if it spreads to Saudi Arabia and the other major oil producers in the Middle East then the result would quickly become—well, that is the worst case, of world recession of serious magnitude."

To what extent, I asked, was a model of this kind an attempt to build not just a workable means of predicting but a formal theory of economics?

"Originally, the first models ever built were to *test* theories in economics," Klein said. "We were all influenced by the theories of John Maynard Keynes—but where Keynes was a back-of-the-envelope man, who thought he had it all in his head, we were trying to introduce a mathematical approach. But in the beginning, in the early forties, all the model-builders were of the left wing of the Democratic party; they were the New Dealers and they wanted to use models for activist policies and to implement policies of the Keynesian type. But as things have developed, over the years, the original group that got it started has drifted away. But economic theory and modeling— First, modeling was to test theories, but now, really, the role of theories is to give us hunches about the way the economic system is put together, so we can try to improve the model. I say that the economy is governed by the big equation system. But one can't see that. So how do we know? You have to go by your sense of the ways human beings interact and the ways institutions interact—and generate an approximation to this reality by continual trial and correction."

The essential point, I said, seemed to be the feedback, the constant checking of predictions against what really happens, so that the model is corrected.

"That is exactly right," Klein said. "We check ourselves every quarter, or even more often."

I said that I imagined the process of improving the model as one of narrowing its performance closer and closer to the track that events themselves seem to be following. Didn't that mean that unusual insights, radical new policy suggestions, were ruled out by the very nature of the modeling process?

"Well, last year there were some radical policy suggestions," Klein said. "*We* didn't make them, but— You see, there are a lot of people outside the academic community working on econometric modeling, too, and it's becoming a big business. I regard our operation here as, among other things, trying to keep people honest. You see, there is a great danger, in working with the complicated equation-systems, that they become sort of a black box. They are locked in the computer. The outsider can't see them; the insider can then shift things around to get a favorable result. So it's very important that all the details be available. Our models are published—in the public domain."

Where is econometric modeling headed?

"The first thing that's happening—and it's on a very, very big scale—is a technology transfer. Modeling will grow very fast in Europe and then in the developing world. At the present time, at Wharton Econometrics, we have two trainees from the Soviet Union, several from Latin America, one from Germany working with us, and one from Japan. And that's just at this moment. Over the years, at any given time, there are four or five people with us from the developing world or from Europe."

How fast is the Soviet Union picking up economic model-building?

"Very fast," Klein said. "And the United States government has always said, 'We want you to interact with Eastern Europe and the Soviets. We don't care if you show them our models; we just want them to think along engineering and modeling lines, so that there will be more rational ways of interpreting each other's actions.' In fact, the United States would like to see centrally planned economies using, as much as possible, the same kinds of statistical modeling methods *we* are using. Then there's more order in the dialogue.

"In the central banks around the world, in the ministries of finance and of trade, and in academic centers, this approach is gaining large support. And that is where the next big thrusts are going to be. To internationalize the model—and to automate it. One project I'm proposing, with a colleague, is to begin with a common data base, worldwide, in the LINK system." LINK is a world-trade model, housed at the University of Pennsylvania, that joins up national models in many countries. "We have the data bases on file here on the computer, and we use a satellite communications network. And in each of several major countries of the world, we hope to put a model-builder, an economist, before a computer terminal. All simultaneously. And we shall have a computer conference, all going to the same data base, and looking at possible policy responses. That is, say, Japan puts a tax on imports—and solve the world model. Germany reacts to that—and solve the world model. The U.S. stimulates its economy—and solve the world model. Play out the actions and the reactions."

Economists all over the world, at their computer consoles: it sounds like an enormous game, I said, Monopoly on a fantastic scale.

"That's it exactly," Klein said. "It is a million-dollar Monopoly game."

How far ahead can the model look?

"The errors, or rather the margin for error, gets bigger as you go farther out," Klein said. "Of necessity, the way the economic decision-making of any large company or any government is run, you have to look ahead one year. That's minimum. You want to get a second or a third year, to know at least if you're going to have a turn—whether things will continue in the same direction, or will turn up or turn down. That's about the range, now. And we do make long-term projections for long-term processes, like the energy process, the forestry process, the life-insurance process—things that take a long time to evolve. But we tell people frankly that although this is our best judgment to the end of the century, say, there is a big error-band. One has to make certain judgments. And one must have a basis for judgment. But one can't make the judgment without taking account of the fact that the band between the upper and lower limits of probable error grows very wide. My main objection to people who have forecast world disaster by the twenty-first century is that the error bands of these predictions are so wide that practically anything can happen and still be within the forecast range. You cannot say with certainty whether the world's ever going to break down, or not."

What was Klein's own hunch?

"My own hunch is that always the doomsday—Repeatedly, the doomsday theorists have been wrong;

and, for example, if you take their models and set them running at the beginning of the *nineteenth* century, and then simulate what happens from then, you would have to predict that our present world could not exist, that everything would have broken down long ago. But we are living proof that it hasn't done so, quite yet."

7

Strong Predictions

If you dig under the north side of that oak tree in the forest (says the gnome in the fairy tale), you will find a jar filled with gold.

Why do we accept a model or find a theory persuasive? We may value science for many sorts of reasons—for the beauty of the patterns it reveals or, at the other extreme, for the power that it ambiguously promises. But much more immediately we accept a model or a theory or an interlocking set of theories because—in a particular and special fashion—it astonishes us. A theory—this is the nature of theories and of models that function as theories—must generate predictions of surprising new facts. Prediction is, of course, the most characteristic and pervasive element in animal behavior and in human speech and thought: it is the dynamo of the feedback cycle. So it's inevitable and yet pleasing that of all the elements and activities that make up the enterprise of science, prediction is the hinge on which the rest swings. And the special quality of the predictions that matter most in science is that they are *strong* predictions. Successful models, successful theories, tell just where to dig for treasure that is dramatic and unexpected.

The reception of Linus Pauling's model of the alpha helix provides an amusing case study. Announcement of the model caused tremendous excitement among scientists—excitement and in some laboratories chagrin. At the one rival laboratory where chagrin ran highest, Pauling's model provoked the discovery of a new fact within a few hours. The story was told to me years later by the man who made the observation, Max Perutz.

In Cambridge, England, at the Cavendish Laboratory, a group led by Perutz and backed by the Cavendish Professor of Experimental Physics, Sir Lawrence Bragg, had been investigating protein structures for more than a decade. Bragg was several years older than Pauling. He had won his Nobel Prize in 1915, in physics, at the age of twenty-five (still the record), for the discovery of that essential technique, X-ray crystallography, by which the three-dimensional structures of molecules can be determined. Using the method, Bragg in 1912 had first solved the structure of one of the simplest possible molecules, the crystal of sodium chloride—common salt. Since the late twenties, Bragg and Pauling had kept up a running rivalry to solve increasingly more difficult molecular structures. Proteins were thousands of times larger and more complicated than any structures previously gotten out. Pauling had been working up by way of the structures of the components—the amino acids, and the bonds that link them in the protein chain. Bragg thought there might be a shortcut. With Perutz and another colleague, John Kendrew, he catalogued many possible models for protein chains that more or less fit the available X-ray evidence, actually building several of the plausible models; in the spring of 1950, the three published a long paper about them. Somewhat uncertainly, they voted one model the most likely to be found in proteins. But they got it wrong—and in part because they failed to follow the rule already established by Pauling that the bond linking two amino acids must be flat.

Pauling and Robert Corey first published the full details of the alpha helix in the May 1951 issue of *Proceedings of the National Academy of Sciences* of the United States. Simultaneously, they published six more papers on other aspects of protein structure. Perutz first read through the Pauling bonanza on a Saturday morning at the Cavendish. He saw at once that the alpha helix was the crucial idea—and, from his own thorough knowledge of the evidence about protein structures, that it should be right. Perutz also noticed that one measurement that could be made on the model showed that the rise from each amino acid to the next, up the helix, was 1.5 angstrom units—which is six-billionths of an inch, truly a tiny spacing even in molecular terms. But if that was correct, and if the alpha helix was indeed present in proteins, then X-ray patterns from proteins should show a characteristic spot caused by that spacing within the crystals. The spot had never been observed—but Perutz knew that such a fine spacing lay at the extreme limit of resolution of the X-ray method. So after lunch Perutz devised a novel experimental setup. He took a single horsehair, because hair is made up almost purely of the fibrous protein called alpha-keratin, and placed it at a calculated angle down the center of a cylindrical sheet of film, in order to catch spots that would otherwise be missed. He shot a single X-ray picture. In the pattern was the predicted spot. On Monday morning, he showed the pattern to Bragg, saying that Pauling's triumph had made him so angry at their collective stupidity at the Cavendish that he had had to check the structure at once. Bragg said only, "Perutz, I wish we had made you angry sooner!"

Pauling's model made a strong prediction. It told Perutz precisely where to look to find a phenomenon that had never been seen or expected. And the story has something more to say about how predictions work in science. Because the group at the Cavendish was in direct competition with Pauling, they were the best informed and equipped to note the prediction and to verify it—and simultaneously the most ready to point out any failure. Indeed, two years later Pauling and Corey built and published a model of another biological substance, deoxyribonucleic acid—DNA, the stuff that genes are made of; but in this case they were wildly wrong, and scientists at the Cavendish leapt upon the error with savage glee. The creative dynamism of science arises from the fact that those to whom its latest predictions mean the most include exactly those who are most disposed to be critical. The predictions of astrologers astonish believers. The strong predictions of science astonish informed skeptics.

We make predictions literally all the time, and survive only by their accuracy. Our slightest automatic movement—reaching for a piece of food, glancing up to see what made the unexpected noise—is possible only because we can make that continual alternation of tiny predictions and corrections to converge on the target. Speech, conscious thought, our experience of making choices, all reflect the processes by which the brain uses the past to predict what may be coming. "Sufficient unto the day is the evil thereof" and "Don't borrow trouble"—but, incorrigibly, man binds time. We bind the future with the past even in sleep. As far back as we know, people have thought that dreams somehow predict the future; Sigmund Freud speculated that dreams, instead, replay the individual's past, both immediate and childish; but investigators of sleep and dreaming have recently concluded that dreams reflect a preparing for tomorrow—that they are predictions after all, but only in the most prosaic way, fragmented sentences from our rehearsals for the next day's problems. Recall George Herbert Mead's aphoristic definition: "Value: the future character of the object insofar as it determines your action to it." Dreams are colorful sharp splinters of the unceasing process of valuing. Mankind, desperate to predict, has put faith in means more bizarre than dreams. We have strained to interpret the mumblings of Apollo's priestess, intoxicated with the smoke of smoldering ivy leaves—and prophecies that could mean anything at all are to this day called Delphic. We have tallied the almost perfectly random permutations of forty-nine yarrow stalks in order to pick out our doom from the almost perfectly ambiguous paragraphs of an ancient Chinese book, the *I Ching*. We have supposed that the incalculable vagaries of personal fortunes depend upon the entirely calculable and ineffably remote paths and patterns of the planets. And now we listen in delicious dread to scientists' predictions, say, that the starry universe will collapse into itself and vanish in an anti-big anti-bang, twelve billion years hence. Or that the earth's climate is getting colder (or, by another authority, warmer) at such a rate that our children will suffer a new ice age (or, as it may be, suffer the melting of the ice caps and the drowning of New York, London, Tokyo, and Sydney). Even when vague, ambiguous, susceptible of multiple interpretation, and impregnable to checking (twelve billion years?), predictions have enormous rhetorical force.

Eventually, of course, the value of a theory, its rightness and explanatory power, has nothing to do with its psychological impact. Nonetheless, persuasion is certainly one function, an entirely legitimate and necessary function, of the strong predictions of a scientist's model or theory. The strong predictions most astonish the most informed skeptics—and set their minds to working out the next consequences. Strong predictions force the growth of theories. They turn hypotheses into research programs.

Edmund Halley was an English astronomer and mathematician, a friend of Isaac Newton's. Halley learned of Newton's theory of gravitation and pursuaded him to publish it and the resulting calculations of the elliptical orbits of the planets and the perturbations they exerted on each other: this became Newton's *Principia Mathematica,* which Halley in fact financed and helped prepare for the press, and which appeared in 1687. No previous theory of the motions of the planets and stars—not the tattered ancient Ptolemaic earth-centered system, not Copernicus's heliocentric circles nor Kepler's ellipses—had made the appearance of comets an integral part of the account. Superstition held comets to be warnings of vast disasters. Kepler had suggested that comets were heavenly bodies that moved in straight lines—but this was hardly better than saying they were anomalous and unpredictable, which the superstitious knew already. But by Newton's theory, some comets should swing past the sun in open curves—hyperbolas and parabolas—and so would never return, but others should move in ellipses like the planets' except much longer. A spectacular comet had dominated the sky for months in 1682; Halley had made repeated sightings of its path, and now put Newton's method to work on that. Although the observations, of course, covered only a small segment of the complete orbit, Halley calculated that the comet reappeared every seventy-six years. He looked back into astronomical records and found that, sure enough, a great comet had appeared in 1607 and one in 1531. Halley's calculations, and those appearances of his comet, had formed no part of Newton's original theorizing: to elevate those unaccountable previous incidents into full and consequential illustrations of the new theory was itself a strong instance of what has recently been called, if not prediction, "postdiction." Then Halley figured to the very night when the comet should appear again and in what quarter of the sky. Newton died in 1727, Halley in 1742; Halley's comet reappeared in 1758, on the night and in the quarter. The case is spectacular, well known, and in principle exactly like Max Perutz's story of first confirming Pauling's alpha helix.

Astronomy and physics have provided the classic examples of model-making and theory-building for historians of science—particularly, indeed, that tumultuous two centuries of astronomy that saw the Copernican model of the heavens succeed the Ptolemaic, only to be superseded by Kepler and then by Newton. Sometimes in that series the newer theory was more accurate than the predecessor, sometimes simpler, sometimes more elegant or more comprehensive—but recent careful reconstructions of the actual content of those theories at their crucial confrontations have shown that none of these advantages was necessarily or automatically on the side of the new. The last swords forged by the craftsmen of the age of bronze were finer, stronger, sharper, after all, than the first swords of iron. What made the difference at each stage, though, was that the newer theory or program of theories predicted new facts—facts unexpected or even contradicted by the earlier or rival programs.

The long, peaceful reign of Newtonian theory which followed was punctuated by strong predictions—of which the most astounding (and instructive about prediction's real role) led to the discovery of the planet Neptune. When Newton first published the *Principia*, the details of his theory were still so crudely worked out that everyone who was equipped to follow it saw clearly that it could not satisfactorily account even for the orbit of the moon. But by the nineteenth century, the gravitational interactions of the planets were understood, in Newtonian theory, with great and reassuring accuracy—and one awkward exception. The planet Uranus—itself discovered by telescope in 1781, by Sir William Herschel, who first thought he had found a comet—had an orbit that failed to conform exactly to calculations. The observations that showed this were abundant by the 1840s. The anomaly was perhaps not felt as a threat to Newtonian theory, but it demanded explanation. The English astronomer John C. Adams and his French colleague Urbain J-J. Leverrier therefore invented a new planet, never yet observed, moving in a farther orbit and perturbing Uranus's path. By Newtonian methods, they figured their new planet's position. Leverrier sent the calculations to Johann G. Galle, an astronomer at an observatory in Berlin. The night after Galle received the letter, he aimed his telescope where he had been instructed and found the planet.

An inseparable aspect of Newtonian theory was the Newtonian mathematics—the calculus, which he developed as the language of his celestial mechanics. Mathematics, of course, pervades the physical sciences; its power appears most vividly in the process of prediction. Mathematics says nearly nothing, in the poet's sense: the interplay of multiple local meanings and resonances that gives ordinary language its richness, warmth, and layers of meaning is here deliberately suppressed. The sonnet sequences of the mathematician are sometimes beautiful; they are always austere. Sharply, narrowly focused, mathematics reaches far. Abstract, formal, it leaches out the similar skeletal patterns to bring together apparently dissimilar phenomena. The physicist John Ziman wrote recently, "The essence of mathematical reasoning is that it is perfectly transparent in thin sections, yet intellectually opaque in bulk." Though each little step is clear and compelling, where those steps may lead—particularly now when big computers can pace off an almost unimaginable number of them—cannot be fully anticipated even by the adept. Thus, by reaching far and by unifying, mathematics can build patterns, models, theories that are fertile beyond the ordinary in predicting unexpected new facts. And the predictions are unusually free* of ambiguity; their very precision facilitates verification and makes success the more astonishing.

After Uranus generated Neptune, an anomaly among the orbits of the planets still remained. Mercury, the innermost, traverses its ellipse around the sun once every 88 of our earthly days; at the point on the ellipse closest in—called perihelion—the planet is 28.5 million miles from the center of the sun, all but skimming the surface, and has slid far down the hill of the sun's gravitational influence and is traveling very fast.

*Not perfectly free of ambiguity, necessarily: the history of mathematics is full of important new examinations of supposedly settled problems, and these innovations have typically begun with the perception that an old theorem contains new, additional meanings. Mathematics itself proceeds like an experimental science.

But the perihelion of Mercury changes position. It advances slightly at each full rotation. The entire ellipse slowly walks around the sun. Newtonian calculations of the gravitational influence exerted on Mercury by the other inner planets certainly accounted for most of the shift. But in 1845, the year before predicting Neptune, Leverrier refined the Mercurial calculations and discovered that a portion of the advance of the perihelion stubbornly refused to be charged to the other planets. Mercury's perihelion advanced too much by 43 seconds of arc per century, or a degree every 837 years. The response was the standard Newtonian defense: Leverrier suggested that Mercury was additionally affected by an unknown planet—though where it lurked was problematic, with Mercury relatively close to both earth and sun, and moving fast. The perturbative planet was named Vulcan. It eluded observation.

In 1905, Albert Einstein published his first paper on the theory of relativity. This introduced what he later called the special theory of relativity—special because it did not apply to all types of relative motion, but only to the special case of things in motion with constant velocity relative to each other, not accelerating or decelerating. One consequence of the theory of relativity was an excited burst of work and publication as other scientists rushed to exploit the new problems that relativity opened up. Einstein has conventionally been pictured as somewhat isolated from the scientific community; as everyone knows, he was a minor bureaucrat in the Swiss federal patent office in Bern when he published the special theory of relativity. But after 1905, Einstein and the problems he pursued were in the central line of the enterprise of physics. Relativity was not only a theory, it was a research program.

Over the next decade, he and others made repeated attempts to understand how the principle of relativity could be given a general form that would include acceleration—and most importantly the acceleration of gravitation. The crucial and difficult new idea Einstein developed from a suggestion made in 1907 by Hermann Minkowski, a mathematician who had been one of his teachers. The idea was geometric—that if space and time are thought of as being a four-dimensional continuum (Minkowski's notion), then the equations of relativity make gravitation itself emerge as an expression of a curvature, or warping, introduced into the geometry of space-time by the presence of massy matter. The idea seemed remote from common sense (as did in their day, of course, the idea that the earth spins, that the planets go around the sun, that germs cause disease, and so on). Einstein himself found the idea difficult; he took several steps to elaborate it into a general statement. One landmark was a paper in 1911; he introduced the general theory of relativity in papers in 1915; he published the fully worked-out treatment in 1916.

None of the work through the first paper of 1915 was formulated with any thought or reference to the perihelion of Mercury. One unforeseen consequence of the general theory was immediately pointed out by the physicist Karl Schwarzschild. If the orbit of Mercury was recalculated by the Einsteinian geometry, the perihelion advanced more quickly than the Newtonian dynamics had allowed. The difference was slight, being exactly 43 seconds of arc per century.

Was this the prediction of a remarkable new fact? The irreducible disagreement between observation and Newtonian calculation had been known for seventy years or more. Dramatic postdiction, then—for the fact that had been hardly more than a persistent minor irritation to the old theory found its significance transformed into the unexpected, direct, and integral consequence of the new.

General relativity forced such a wrenching readjustment of patterns of thought that for years it met hostility, even from many scientists. The skepticism was healthy, for it meant that relativity has generated more exacting tests than any other theory or research program except perhaps Newton's own. One prediction from general relativity was particularly strong: not only matter but light should respond to gravity in a way determined by the effect of mass on the geometry of space-time. Newton himself and his immediate followers had thought that light is made of particles. Applying Newtonian gravitational mechanics to a beam of such particles indeed suggests that gravitational effects ought to be observable, though nobody had ever looked for them. The alternative theory, already persuasively argued in Newton's day by Christian Huygens, a Dutch physicist, is that light consists of waves—and in the nineteenth century new evidence for the wave theory had mounted overwhelmingly. Einstein's contemporaries were shocked by his prediction about gravitation and light. He first made it in the paper of 1911, before he had worked out the general theory. There he calculated from special relativity alone that a ray of starlight that passed close to the sun would be bent: the star would appear to an earthly observer to be displaced from its usual position. And he wrote, "As the fixed stars in the parts of the sky near the sun are visible during total eclipses of the sun, this consequence of the theory may be compared with experience." But the displacement he predicted in that paper was .83 of a second of arc, which was close to what Newtonian mechanics could be made to produce. When Einstein got the geometry of gravitation worked out in the general theory of relativity, he refigured the effect of mass on light—and in the comprehensive paper of 1916, putting the sun's mass into his new equations, he generated the prediction that "a ray of light going past the sun undergoes a deflection of 1.7 seconds of arc."

The English astronomer Sir Arthur Eddington

had been a proponent of Einstein's theories from the beginning. Toward the end of the First World War, British astronomers noticed that a total eclipse of the sun would sweep across the South Atlantic on May 29, 1919—a remarkably favorable eclipse, too, because on that day the sun would have as its backdrop a patch of sky exceptionally rich in bright stars. When the war ended, Eddington and others organized two expeditions by the Royal Astronomical Society—one to Sobral, in Brazil, and the other, led by Eddington himself with another astronomer, to the Portuguese island of Principe, off the coast of West Africa. Several years later, Eddington described the experience on Principe in a popular book about relativity:

> On the day of the eclipse the weather was unfavourable. When totality began the dark disc of the moon surrounded by the corona was visible through cloud, much as the moon often appears through cloud on a night when no stars can be seen. There was nothing for it but to carry out the arranged programme and hope for the best. One observer was kept occupied changing the plates in rapid succession, whilst the other gave the exposures of the required length with a screen held in front of the object-glass to avoid shaking the telescope in any way.
>
> *For in and out, above, about, below*
> *'Tis nothing but a Magic* Shadow-*show*
> *Played in a Box whose candle is the Sun*
> *Round which we Phantom Figures come and go.*
>
> Our shadow-box takes up all our attention. There is a marvelous spectacle above, and, as the photographs afterwards revealed, a wonderful prominence-flame is poised a hundred thousand miles above the surface of the sun. We have no time to snatch a glance at it. We are conscious only of the weird half-light of the landscape and the hush of nature, broken by the calls of the observers, and the beat of the metronome ticking out the 302 seconds of totality.

They had time to make sixteen photographs, with exposures from two to twenty seconds. Only one was usable. It showed "fairly good images of five stars." They had brought with them photographs of that same patch of sky, taken through the same telescope in England in January, when of course the sun was out of the way. Now before leaving Principe they placed the eclipse photograph and a comparison photograph one atop the other in a light box so that corresponding images appeared close together. The measurements showed the five stars each displaced by 1.61 seconds of arc—and with a margin of possible experimental error that was larger than the difference between the observation and the 1.7 seconds of the theory's prediction. The results were presented in London the following November, to a joint meeting of the Royal Society and the Royal Astronomical Society. A portrait of Newton looked down upon the crowded, hushed audience. And by the end of September of 1919, a physicist friend of Einstein's had cabled him Eddington's result. Einstein, in Berlin, wrote a postcard to his mother, in Switzerland, beginning, "Joyful news today. H. A. Lorentz has telegraphed me that the English expedition has really proved the deflection of light by the sun."

The success of the physical sciences has made their methods and the mathematics that is its essence a seductive model for other sciences. Social scientists, in particular, have long been struggling to turn their disciplines mathematical. The first difficulty is that the parameters of human behavior are many more, and more elusive, than those of physics. But setting that aside, the model of the physical sciences leaves out at least the possibility—understood by a few of the great sociologists—that the social sciences at their best are hybrids, requiring some of the methods of scientists, perhaps, but surely, as well, the scholarship and flair of the historian. And the effort to be mathematical can confuse the tool with the aim. The aim is a thriving line of research with theories that have scope and coherence, that grow—and that throw off predictions of unexpected new facts. Here the record of the social sciences is thin.

Already in the biological sciences the power of mathematics wanes. To be sure, it's often essential that experimental results be expressible quantitatively, and that the anticipated consequences of a new idea be stated precisely. Occasionally, some highly physical aspect of a biological problem permits an intense and elegant mathematical attack—as in the X-ray analysis of the helical molecules typical of living processes, or in the calculation of the transmission of electrical impulses along nerve fibers. But in biology for the most part—and compared to the inclusive grandeur of astronomy and physics—predictions have a less formal theoretical base, are limited in scope to particular problems, and offer neither ideas nor data where powerful mathematical methods of analysis can grab hold.

Yet some predictions in biology have been spectacular. To my mind, the greatest was made by William Harvey, the English physician and anatomist of the seventeenth century who first understood the circulation of the blood. Today we take that circulation not as theory but as fact. In the centuries before the microscope, before anesthesia made possible experiments with live animals, and when the use of human corpses for anatomical investigation was set about with prohibitions, it was genuinely difficult to demonstrate directly the many different facts that must be understood in order to trace the full cycle of the circulation of the blood. Harvey had to show that the arteries and veins carry nothing but blood—no streams of air or of bile or of other substances. He had to establish that the heart is an active pump—for example, that the beat we hear and feel occurs when the heart contracts, not when it expands, and that the valves between the chambers of the heart impose a one-way flow. He had to bring into one scheme the two aspects of the movement of the blood—the major movement from the heart through the body (and back), the smaller loop from the heart to the lungs (and back). And for his vision of the circulation to be possible at all, Harvey had to assert that the blood pushed out through the arteries to the tiniest visible arterioles *and then made its way through the tissues* to the venules and so through the larger veins back to the heart. His theory had to postulate the existence of connecting vessels, which he said were invisibly small—a new fact, a new anatomical feature, for which neither he nor anybody else had direct evidence. Harvey published his theory in the celebrated treatise *De Motu Cordis et Sanguinis* ("On the Motion of the Heart and Blood") in 1628. In 1661, Marcello Malpighi, an Italian physician and one of the first microscopists, opened the chest of a living frog, partly dissected and lifted out a lung, placed it on a glass plate (illuminated from below, through a tube by a lighted candle), and examined it with a two-lens microscope he had made. He saw minute vessels in the lung tissue that were carrying blood from the network of arteries to the network of veins. Malpighi named the connecting vessels capillaries. In 1673, the Dutch microscopist Anton van Leeuwenhoek reported, in a letter to the Royal Society in London, that he had examined a live tadpole under the microscope. The tail of the creature being nearly transparent, Leeuwenhoek had seen blood moving from the tiniest arteries through to the veins—had seen, in fact, individual blood cells folding over to fit through the connecting vessels. Capillaries existed in the tissues, too.

Quantitative reasoning played only a small part in Harvey's design—no more, really, than the crude observation that the amount of blood in the body and the rate of its pumping through the heart required that it be used over and over. What mattered, first, was the imaginative breadth by which Harvey saw a unifying dynamic pattern in what had been a confused jumble of facts and claims and traditions—and then the ruthlessness with which he threw out previous anatomical notions, some because he could demonstrate that they were indeed wrong, some because his theory told him they must not be right.

Even in physics and chemistry, as Mendeleev bracingly reminds us, pattern may well precede mathematics. It's cautionary, too, that the science where the power of mathematical prediction was triumphant for centuries, astronomy, finds itself—at the moment, anyway—in a fiercely frustrating condition where richly interesting new observations and clever ideas abound yet large-scale theory and prediction are stuck. A while ago, in a conversation with Philip Morrison, a theoretical physicist at the Massachusetts Institute of Technology, I asked about that. "Astronomers these days predict hardly anything," Morrison said gloomily. "It's a bad time for theory. We predict nothing—just rationalize the observations afterwards." Did he have an example? Morrison made a sour face, then said, "Everyone believes that pulsars are rotating neutron stars. And that idea was put forward the same week the observation was reported—maybe it was six weeks before or after. And from that idea, you can write a page or two of equations that define the model and are roughly correct. But unfortunately we have not been able to get beyond that in all the ten years since."

A Conversation with Karen McNally

"Predicting an earthquake in somebody else's country . . . is something you do with great trepidation."

Karen McNally is an experimental seismologist: she is trying to model earthquakes in the laboratory and to see whether the predictions that follow from the model are borne out. On August 29, 1978, she was visiting the Institute of Geophysics at the University of Mexico, in Mexico City. She had been invited by the director of the Institute, Dr. Ignacio Galindo, and his colleague Professor Lautaro Ponce, to give some lectures on field studies that she had been conducting in Southern California—she was then a senior research fellow at the California Institute of Technology—of the patterns of swarms of tiny shocks that precede larger earthquakes. That evening at the Institute of Geophysics, alarm bells began to ring. The seismographs in the basement were registering something significant. McNally followed her hosts as they rushed to the instruments. The pens on the recording devices were still jiggling; a first calculation showed that the earthquake was taking place in the southeast, about 300 miles away, on Mexico's Pacific coastline in the state of Oaxaca. The earthquake was of moderate force, centered under a stretch of the coastline that was known to seismologists as the Oaxaca gap—which made it of the highest interest, for the appearance of any earthquake activity in the Oaxaca gap suggested that one of the strong predictions of modern geophysical theory was under test. McNally found herself in the middle of a scientific controversy and a delicate problem of international relations. Three months later, she and her Mexican colleagues trapped a very large earthquake.

The Pacific coastline of southern Mexico lies on a known geological fault—which is to say that it's a region of frequent earthquakes, little and big. The presently accepted model of the earth, called the plate-tectonic model, pictures the earth's crust made up of a number of large, thick slabs, or plates, floating on a denser, semi-molten material beneath. The plates are moving relative to one another, propelled by such forces as the upwelling of molten matter along giant cracks at the bottom of the centers of the oceans, forcing the sea floors to spread. The continents ride on some of the plates. Earthquake zones, the theory says, are particularly likely to appear where two plates moving in different directions collide. Sometimes the plates are grinding past each other, which creates what is called a strike-slip fault; the notorious San Andreas fault and the network of related faults in California are primarily strike-slip. Sometimes the collision is head-on, which forces one of the plates to dive beneath the other in what is called a subduction zone, and which creates what is called a thrust fault. The diving plate lifts the other plate—the process that built the Himalayas as India collided with Asia, and that is building the Andes. The southern coastline of Mexico is believed to lie on a thrust fault where the Cocos sea plate dives beneath the Americas continental plate; the coastline is mountainous.

Sometimes, the model says, two plates in collision may slide relatively smoothly, the force of their meeting dissipated only in frequent small earthquakes. But commonly two plates lock against each other and move mainly by fracture, in large earthquakes. Plate-tectonic theory suggests that, over many decades, movement occurs along the entire length of such a plate boundary at a characteristic average rate. If a long time has passed since a particular segment of that boundary has slipped in a large earthquake, that segment is called a seismic gap—here, a gap chiefly in the spatial sense. A seismic gap is an ominous sign: at the lock, enormous energies may be building in unrelieved strain. Such ideas were a developing part of plate tectonics in the early 1970s; their application to Central America was pointed out in 1973, by John Kelleher and two colleagues at the Lamont-Doherty

Geological Observatory in New York, in a paper provocatively titled "Possible Criteria for Predicting Earthquake Locations and Their Application to Major Plate Boundaries of the Pacific and the Caribbean."

A relatively new detail of the model says that at such a locked portion of a fault, not long before a big earthquake even small shocks may stop completely—and even though earthquakes continue to occur at other places along the fault. Here the model presents the seismic gap—the absence of earthquakes where and when they would be expected—chiefly in the temporal sense. The quiescence, called the alpha phase, may last for several years. Then in a beta phase a number of small foreshocks occurs, perhaps over many months or in no more than a few days. In 1977, Gary Latham and two Japanese colleagues, Masakazu Ohtake and Tosimatu Matumoto, at the Geophysics Laboratory of the University of Texas in Galveston, put worldwide seismographical data onto a computer and then searched for temporal seismicity gaps in likely earthquake zones along the rim of the Pacific. They found such a gap in Mexico, a stretch of about 300 kilometers along the south coast in the state of Oaxaca. The title of their paper called the gap a "probable precursor to a large earthquake."

Appearing in seismologists' journals, these bold claims were somewhat shielded; professional readers no doubt realized, if they thought about the matter, that in fact two completely different kinds of predictions were being made. The first was a prediction of great but simple public interest, like forecasting a hurricane: an earthquake will soon occur along this stretch of the Mexican coast. Such a prediction had standing if the theory was taken as proved. The second was a prediction of a strictly theory-making kind that said, on the contrary, that the model is up for testing. In the absence of plate-tectonic theory, about all that had been known with confidence was that earthquakes mostly occur where they have occurred before, so that a seismicity gap should mean, if anything, that earthquakes were less likely there than elsewhere along the fault. But if the detail of the plate-tectonic model that ascribes observed gaps to locks under strain is correct, then a violent shock should be building. This was a strong prediction—the sort that makes a difference for a theory's plausibility. It was made yet stronger by the specific assertion that before a big quake seismic activity would disappear completely for a while in an alpha phase, and would then show premonitory foreshocks.

The group at the University of Texas noted that in the middle of 1975 earthquakes in the Oaxaca gap large enough to be detected by the worldwide network of observational stations—meaning earthquakes of the moderate level of about 4.0, or larger, on the Richter scale—had suddenly ceased altogether. They made a very precise prediction of the place where they thought a large temblor should occur: between 16° and 17° of north latitude and between 96° and 97° of west longitude, which was on the coastline at the center of the gap. They made a very precise prediction of the force of the temblor: between 7.25 and 7.75 in magnitude. For comparison, it was a pair of earthquakes of magnitude 8.2 and 7.9 in a sixteen-hour period that devastated Tangshan in China in 1976, killing at least 650,000 people. The group at Texas could make no prediction of the date of the temblor, but said that the gap had already lasted "substantially longer" than similar periods earlier before large quakes in the region. The last of those had been in 1968, to the west of the present gap, and had badly damaged Oaxaca's capital town, Pinotepa Nacional. They wanted to stake out the Oaxaca gap with sophisticated instruments of all sorts. They applied to the United States Geological Survey for a grant of $289,000 for the work. They were turned down.

The forecast soon became more than a matter for professional argument. Early in February of 1978, the President of Mexico got a letter from two people in Las Vegas, Nevada, claiming that a large earthquake, with flooding, would hit Pinotepa Nacional on April 23. A copy of the letter reached the mayor of Pinotepa. Local people were terrified; many sold their houses for whatever price they could get, and speculators moved in. Then, early in April, United Press International carried a story about the prediction that had been made at the University of Texas, and Mexican papers picked it up. TEXAS U. PREDICTS BIG MEXICO QUAKE said a front-page headline in Mexico City. The UPI account was closer to the original scientific paper, and mentioned no date, but it appeared to give legitimacy to the first reports. Widespread panic followed. The mayor of Pinotepa later said that the prediction in 1978 had caused more damage to the town than the earthquake of 1968. A newspaper up the coast in Acapulco proclaimed that half-a-dozen nuclear charges were buried in a fault off the Oaxaca coast, to be detonated on April 23; others thought that oil or uranium had been discovered and that foreigners had created the panic so that they could buy land cheaply. On April 23, the governor of the state of Oaxaca came to Pinotepa, a visit that had been announced to reassure the local people; festivities had been organized, including folk-dancing groups and music and a display of disaster-relief methods; nonetheless, at least a fifth of the populace were out of town. At 5:40 that afternoon, as it happened, a small earthquake—nearby, but not in the Oaxaca gap—rattled the doors of the town hall.

Karen McNally's chief interest lies at still another level of theory and prediction in seismology. Plate tectonics models a mechanism for the accumulation of strain before a major earthquake. She wants to understand in fine detail how that strain is discharged—the pattern of what seismologists call "pre-failure" leading up to a big earthquake. Ideally, she would like to compare detailed analysis of patterns of foreshocks, measured in the field in the weeks and days before a large quake, with a similar analysis of the accumulation and relief of strain in small samples of materials under great and increasing pressure, controlled and measured in the laboratory. The material she uses in the lab is not rock but ceramic: hard-fired clay, in cubes measuring a few centimeters on each side, squeezed at great pressures, and measured for changes in volume, in electrical conductivity, and in acoustic—sound-transmitting—properties, up to collapse of the chip. In effect, as the chip is squeezed, before it collapses it squeaks—though at very high frequencies. The cubes of ceramic provide a model of processes that occupy cubic kilometers deep in the earth.

The evening of August 29, when the seismographs at the University of Mexico detected the moderate earthquake in the Oaxaca gap, the Institute of Geophysics was bitterly divided about the predictions—as McNally knew. "Basically, predicting an earthquake in somebody else's country is not—is something you do with great trepidation," she said to me the following spring. Several seismologists at the Institute of Geophysics had been angered by the Texas prediction, which they thought had been irresponsible and to blame for the public panic. They had analyzed the data used by the group at Texas and had proved, they claimed, that the prediction itself was statistical nonsense. Nonetheless, McNally told Professor Ponce that the quake that night could well be the onset of the beta phase, weeks or days before a main temblor. She then was told that five weeks earlier, on July 16, there had been a shock of like size in the gap.

"We didn't have the luxury of *time*," she said in our conversation. "It was clear, that night, that things were going on in that area," she continued. "And I told them, 'You know, I think that the data are good enough. Whether there really is a gap or not, it's certainly one of the better-documented areas for study. We can always go back and rework, reargue the data—but we might never have another opportunity to go out with the instruments.' They didn't have enough people, or enough money. So I came back to Caltech and got a small amount of money and a couple of people and radios and other supplies so we could go out to the area." The two institutes pooled resources; the Mexicans had some new portable seismographs of a simple, reliable design. "We mobilized as quickly as we could—given the circumstances of two countries and two institutes."

Had they had any idea, I asked McNally, how long it might be before a big quake, if there were one at all?

"Well, there were two cases previously in the area," she said. In 1965, at the east of the present gap, a major earthquake had followed a beta phase that had lasted a couple of years. In 1968, the quake to the west that hit Pinotepa had been preceded by a beta phase that was much shorter, "on the order of days. But my idea was that if we went out we could get baseline data no matter what then happened, and could continue to remonitor the area, using spot checks from time to time to compare with the baseline. We couldn't lose by being early!"

In the data they planned to get, McNally was hoping to find evidence for a particular strong prediction that came out of her laboratory work. Scale effects, she said, are almost unimaginably vast in seismology. Time, to begin with. "People have been around, looking at these events and getting reasonable data, since the early part of this century—here in the United States, anyway," she said. "In China, they have a long history, some of observations with instruments but also through reports and ledgers. But geological processes are so slow that to try to predict, on the basis of ninety years of data even in the best cases—on the geological time scale, that's like trying to predict tomorrow's weather based on *one minute* of observation."

The scale effects of physical size are nearly as difficult. She could not use rock in the laboratory. "You can't take something at one centimeter and expect it to be a scale model of itself at earth-type dimensions of kilometers." Instead, ceramic chips promised to be a good model. "There are small heterogeneities within the ceramic material," McNally said. "There are weaker little particles included within it, and small pores, and mismatches at grain boundaries, which are flaws where failure can start as the strain increases. These can model the heterogeneities within the earth's crust at a very large scale. That's one aspect of the problem. The other aspect is how cracks propagate. How rupture occurs. Cracks grow longer, slowly—leading to failure at the heterogeneities. And we want to know the speed at which this happens and its relation to faults in the earth." In the laboratory, McNally said, she had learned that as stress on the ceramic increased, there was a quiescent period; then, toward the point of failure, there were events that corresponded to foreshocks, which could be measured by changes in the acoustic behavior—the squeaks—of the sample. Then, after the foreshocks, there was again a lessening of activity—but very gradually, in a pattern of slow decline that had not been seen before. Only then came the big break. "I think that if we are going to predict earthquakes, first we are going to have to have a definite model of the physics of what's going on," McNally said. "Nobody has been able to do the scaling problem before, and there's certainly no assurance that I can. But because we can conduct the laboratory experiments, then if we know how to relate them to the earth—then we can make extrapolations.

"For the Oaxaca gap, we had certain intermediate-term data. We knew it was on a plate boundary at a subduction zone. And we knew the dynamics and the rates of subduction, and we expected it to break all along the gap, in time—we knew it was capable of rupture, because there had been a big earthquake on either side of the area, so it wasn't sliding completely aseismically. We had all this intermediate-term information. Then the key thing was the resumption of activity following quiescence in that area. I felt that I didn't have to jeopardize my career by *predicting* a big earthquake, but I could certainly go out and acquire the kind of field data that we need for developing physical models in the laboratory—and, at the same time, there *might* well be an impending earthquake and, if so, we would certainly like to know what went on before it. There were no other data available of this kind."

In other words, I said, all she needed to say publicly was that the gap would be interesting to study no matter what happened.

"Yep," McNally said. "In China, you know, it is acceptable to make an incorrect prediction. Their philosophy for earthquake prediction is that it's better to have a thousand days without an earthquake than one day with no prediction! But here it's not okay to make a mistake. The economic impact is tremendous. We don't have the same kind of social and cultural response to a prediction. Real-estate values go down. Your career plummets if there's no earthquake. One of the earliest predictions was made for Southern California in 1976 by Jim Whitcomb at Caltech"—James Whitcomb had successfully predicted a moderate earthquake in December of 1973, which occurred in the next month; in 1976, he predicted a larger one—"and it didn't come. And meanwhile there was havoc. He was criticized, and there were lawsuits pending, based on real-estate values in the San Fernando area. That's a burning case history in everybody's mind who's working in this subject. Our first guesses are going to have to be very carefully made.

"I like calling them 'special study areas,' because in that sense, I think, one strong lesson is that we must be *active* in pursuing the data to work on these problems, despite the risk in predicting earthquakes," McNally said. In Mexico, they had about ten people, all together. They set up seven stations, one right on top of the place where the Texas group had said the earthquake would be centered, the other six in a partial ring, on the land side, spaced about forty-five kilometers from the center. "We went out the first of November. And it took about three days to find sites and get the first one operating, and then the second and the third. And by the eighth of November all the stations were in place and recording data. A field project is not an easy thing, especially in remote terrain. Along the coast, this is a tourist area; but the minute you go inland you are in dense jungle and highland. And you have to visit each site to collect and change the records of each seismometer." The instruments used sharp styluses to scratch smoked drums, "the oldest form of recording and still one of the very best for field work. But they had to be changed every forty-eight hours."

The instruments were sensitive to shocks of less than 1.0 on the Richter scale—not so noticeable as the vibration from a small car passing in a suburban street when one is indoors. For the first ten days, they picked up only slight background activity. About November 15, larger shocks began to be felt, with one of 3.6. After that peak, though, the foreshock activity decreased in frequency and strength until the twenty-eighth, when there was another burst that, in turn, gradually declined over the following hours. At that time, McNally was on her way to read a paper on her laboratory work to the American Geophysical Union, meeting in San Francisco. Driving in Los Angeles on the twenty-ninth, she heard a bulletin on the radio that an earthquake had hit Mexico. She rushed to her lab, then phoned the airport, and flew directly to Mexico, packing in more instruments for observing the aftermath. On November 29, they had trapped a massive quake. "It was right in the middle of our array—exactly. Right on!" McNally said. Its magnitude was 7.8—a fraction higher than predicted.

As far away as Mexico City, there was damage. Two people there were killed. "That was at a great distance," McNally said. "Mexico City sits on an alluvial basin, which amplifies the ground shaking at longer, slower periods as the waves bounce around. Curiously, there was not terrible damage in the epicentral area, though the Oaxaca earthquake was the same size as the disastrous Tangshan quake, in China. Then, after the quake, the governor of the state of Oaxaca came to visit. He flew in—the first visit he'd ever made to the area—with his press attaché and staff, and saw the field operation that we had set up. After the fact, anyway, our team was seen as a very important event in Mexican scientific history—that the earthquake had been trapped."

For the standing of plate-tectonic theory and the strong prediction that seismicity gaps mean locks that presage major earthquakes, the event was in the scientific aspect a triumph. For Gary Latham, at the University of Texas, the event was a vindication. McNally and Ponce and their colleagues both in the United States and in Mexico had a set of data on a major earthquake—its precursors, the event itself, and the aftermath—that was unique and invaluable.

For example, McNally said, there had been a study done at the Massachusetts Institute of Technology which had seemed to suggest that in the period immediately before earthquakes of magnitude 7 and larger, only in 44 percent of the cases did foreshock sequences show up. But the study had used the worldwide data, and the seismographs in the world array do not catch shocks of less magnitude than about 4, at best. In the Oaxaca data, the foreshock sequence was unmistakable. "Yet in this case the largest event was less than 4, and you never would have seen that in the worldwide data on which the M.I.T. analysis was done," McNally said. "So it may well be that the foreshock pattern is much more systematic—except that until now we've thought that they didn't occur in cases that fell below the cut-off. And it may occur on different kinds of time scales. Because, typically, people have simply asked, 'Oh, what happened in the week or so before the earthquake?' But now we are stretching out these studies to the longest records we can get—which means decades. And it appears that there may be extended foreshock behavior that was not previously identified just by looking at the week before—that goes on at slower rates over longer periods, as well as at lower threshold."

For McNally's modeling of the physics of strain accumulation and failure, the Oaxaca data was significant and encouraging: sure enough, once the foreshocks began they reached a peak, then tapered off gradually before the big quake. At least in a preliminary and qualitative way, this pattern matched nicely the peak-and-slow-decline of the squeaks emitted by ceramic chips under pressure, just before they collapsed.

How much had the entire Oaxaca operation cost?

"In Mexico, you can get by quite cheaply," McNally said. "We had rooms in a nice villa in the area for three dollars a night, and food was only a dollar or a dollar-fifty a day. Air fare was two hundred and fifty dollars, round trip, from Los Angeles. Including several trips for analysis of data, and then sending extra engineers and other people down there after the big earthquake with new instruments and to make repairs, and everything—I think we spent about seven thousand dollars. Bear in mind that the Mexicans spent an equivalent amount."

A Conversation with Glenn Seaborg and David Morrissey

The interplay of pattern, model, and prediction in the search for elements beyond Mendeleev.

Among the most wonderful aspects of Dmitri Mendeleev's perception of the underlying pattern of the elements is its vitality—for the periodic table is one of the grand simplifying ideas in science that have had consequences that go far beyond what their originators ever imagined. The consequences have been developed in a balletic interplay of patterns, models, predictions. From the periodic table, Mendeleev correctly predicted the existence of elements that had not been known before. Half a century later, a new model of the atom—embodying ideas that Mendeleev himself had fought against—placed the pattern he had observed on an unassailable theoretical foundation. Today, half a century later again, the periodic table and the predictions made from it are the basis for a search for new elements that Mendeleev would have considered impossible.

Around 1900, after the discovery of the electron and of the phenomenon of natural radioactivity, the evidence grew compelling that the atoms of the chemical elements are not themselves the fundamental, indivisible particles of matter and that, indeed, some elements can decay into other elements. These discoveries were the origin of atomic physics. Yet they seemed to Mendeleev to threaten his periodic law. He denied the discoveries and resisted their interpretation vigorously, to his death, in 1907.

Mendeleev had organized his periodic table in order of increasing atomic weights—a sequence that was hard to test because of the seemingly irregular intervals in it. By 1914, atomic physicists had built a simple mental picture of what an atom is like. Experiments by Ernest Rutherford, models suggested by Rutherford and then by the young Niels Bohr, established that every atom has a nucleus, which is extremely small and dense and carries a positive charge, surrounded by electrons, which are extremely light and negatively charged. The number of positive charges on the nucleus is exactly balanced by the number of electrons. Hydrogen has one positive charge (one proton, we now say) and one electron. Helium has two and two, oxygen eight and eight, and so on. The nuclei of atoms heavier than hydrogen also contain neutrons, which have the same mass as protons but carry no charge. Protons plus neutrons make up an atom's weight; the mass of the electrons is negligible in this context.

In 1914, Henry G-J. Moseley, a young British physicist, showed that the sequence of the elements is put in order and explained, more precisely, by the number of positive charges on the atoms' nuclei. From hydrogen, with its one positive charge, to uranium, with 92, the list had places for exactly 92 elements. Moseley established the periodic pattern on a relationship of extreme simplicity—for it also became clear that the chemical behavior of an element was determined by the electrons it carried, which of course equaled the number of positive charges on its nucleus.

A general model of the atom was then developed. Element by element, down the table, as protons and electrons were added the electrons formed shells. After each electron shell was filled up, a new shell started. Only in the outermost shell were the electrons—or the empty spaces for them, the holes—available to interact with electrons or spaces of other atoms; thus, the number of electrons and holes in the outermost shell gives each element its particular chemical character. The model at last explained the cyclic repetition of chemical characteristics down the table: as successive electron shells were occupied to the same extent, similar chemistry resulted. For example, certain gases—helium, neon, argon, krypton, and radon—were known to be chemically almost inert; they were called the "noble" gases from their disdain for other elements. This turned out to be because their outer electron shells are completely filled; they have no spare electrons or holes to interact with other atoms.

The model also explained the peculiar chemistry of the rare earths—the set of fifteen elements that stretches from lanthanum, No. 57, through lutetium, No. 71. These were an anomaly in the periodic table, for they are almost identical in chemical behavior; indeed, they are "rare" earths not because they are uncommon in nature but because they were extremely difficult to separate from each other in pure form. Atomic physicists now understood that as protons are added to make the elements from lanthanum onward, the new electrons take places not in the outermost shell but in a shell more deeply buried, where the change in number of electrons doesn't much affect the elements' chemistry. With lutetium, the last of the rare earths, the inner shells are filled; in the next elements, the further new electrons go to an outer shell—and, once more, chemical behavior begins varying periodically.

Finally, the general model of the atom explained that when one element is transmuted to another by radioactive decay, the reason is that the number of protons in the nucleus is altered. Mendeleev's pattern had found its theoretical base: his periodic table was explained by the model of the atom that he had rejected.

Then in 1938 the German physicists Otto Hahn and Fritz Strassman announced that they had bombarded uranium with neutrons in an attempt to add to the nuclei and so create an element beyond uranium. And, in fact, they succeeded in producing traces of something in their target that was not uranium. But several months later Lise Meitner explained that what Hahn and Strassman had done was split the atom—to produce a trace of barium, a fission product, a fragment of uranium with about half the number of protons. This discovery set

Glenn Seaborg, a young American chemist at the University of California in Berkeley, to wondering how the periodic table might be extended by making true transuranium elements.

The first of these was found in 1940, in the lab where Seaborg was working, by two of his colleagues, the physicists Edwin McMillan and Philip Abelson. They detected it in an investigation of the fission of uranium and named it neptunium; its atomic number was 93. Then McMillan, Seaborg, and their co-workers, by bombarding a uranium target in the Berkeley particle accelerator, quickly made the next element, 94. They named it plutonium. Plutonium itself undergoes fission when hit by slow neutrons, releasing great energy—and more slow neutrons, generating a chain reaction. The stuff could be used for a bomb. Seaborg went to the University of Chicago, joining the Metallurgical Laboratory—for that was the cover name for the bomb project there—to investigate intensively the difficult chemistry by which plutonium could be purified and handled. Only in 1944 did he get back to the pursuit of transuranium elements; he then manufactured No. 96, curium, and then 95, americium. All told, fourteen transuranium elements have been made to date; Seaborg has had a hand in the discovery of ten. Element 101, obtained in 1955, he named mendelevium. The periodic table has been enlarged through 106, which Seaborg and colleagues got in 1974. (Seaborg was also chairman of the Atomic Energy Commission from 1961 through 1971.)

In the late seventies, Seaborg and a team of young scientists began probing the boundaries of the periodic table in a radically new way. They were testing a prediction that elements much heavier than any that had been imagined before—way out in the neighborhood of No. 114, or 126, or even 164—might be manufactured and might prove stable enough to be detected and identified. One summer, I went to ask him about his fascination with the periodic table. Seaborg's office was on the ground floor of a building at the Lawrence Berkeley Laboratory halfway up a mountain overlooking the university. He had recently turned sixty-six; he was tall, tanned, with thinning white hair—courteous, slow-spoken, and reserved.

"I have always been interested in the periodic table, and how the new elements fit into it," Seaborg said. "Since I was a graduate student, since I was in my twenties. What Mendeleev really revealed when he first recognized the concept of the periodic table—and he based it on the idea of pattern, exclusively—is the repetition of chemical properties of elements, so that they fall into families, in columns, which enabled him to predict the chemical properties of undiscovered elements. The holes in the pattern were the spaces, he felt, that needed to be filled in; and his big contribution was to predict the chemical properties of those missing elements. First, that there *is* an element missing there, and, second, its chemical properties. And he had spectacular success. He just interpolated between the elements above the hole and below it in the column, and came out with predictions that were confirmed within a few years. A lot like a crossword puzzle."

In contrast to Mendeleev's interpolations, "*We* were going *beyond the end* of the periodic table," Seaborg said. "The first new one was discovered here, in Berkeley, in 1940; and then we synthesized the next one, which is 94, plutonium, at the end of 1940 or early in 1941, using the cyclotron on campus."

Had it been predicted that plutonium was going to be a fissionable material?

"It seemed reasonable that it would be," Seaborg said. "And we produced enough, in 1941, to demonstrate that it was fissionable. Just a fraction of a microgram, but it was enough. . . . I was asked to come to Chicago and set up a group to work out the chemical processes for separating plutonium from the uranium and the radioactive fission products produced in the reactor. As soon as we got the separation problem worked out, I could spare a couple of young scientists to try and identify the next two elements, which were elements 95 and 96. Synthesize them by nuclear reactions and, then, chemically identify them.

"Out beyond the end of the table, we were proceeding a little blindly. Until I got the idea that these new elements were part of a new family of rare-earth elements. The previous rare earths, called the lanthanides, beginning with lanthanum, were long known in the periodic table; I called this new family, beginning with actinium, the actinides. And I predicted that the actinides should be analogous, element by element, with the lanthanide series. So that was a matter of pattern, too."

Seaborg's actinide series, the radioactive rare earths, came to an end with element 103. The next, element 104, takes its place in the column of the periodic table beneath hafnium, which is the first after the lanthanide rare earths. Does 104 in fact resemble hafnium, chemically?

"Yes, 104, as far as we know now, has properties similar to hafnium," Seaborg said. To study the chemistry of these elements, though, is difficult. They are produced in minute quantities—and, worse, they decay radioactively in very little time. A sample of 104 is half gone in about one minute, which not only obliges chemists to work fast but presents them with a rapidly increasing contamination by decay products which, of course, have different chemical properties. Element 105 was made in 1970 at Berkeley, and there named hahnium, after Otto Hahn; independently, in the Soviet Union, a team led by Georgii N. Flerov also claimed to have discovered it, and named it nielsbohrium. This one ought to resemble the metal tantalum, No. 73. But it has a half-life of about forty seconds, and no meaningful chemical studies had yet been done. And 106—also discovered in Berkeley and claimed by the group in the Soviet Union, but still not named until the priority is established—has a half-life of about one second.

What about elements beyond 106?

"On the basis of the simplest projections, it would be expected that the half-lives of the elements beyond 106 will get shorter and shorter," Seaborg said. "Down to something like a ten-billionth of a second when you get as far as element 110. That would make the prospects somewhat dismal. But other ideas have entered the picture recently—leading us to be optimistic that we can synthesize and identify elements well beyond the upper limit of the periodic table so far. These have come to be called the 'superheavy' elements."

The attempts to make new elements are carried out at a linear accelerator called Super HILAC, at the Berkeley laboratory; time for the experiments must be reserved months in advance. There, uranium is bombarded not just with neutrons but with projectiles that are themselves the nuclei, stripped of electrons, of middleweight elements like xenon, 54. The chemistry is done in a laboratory upstairs from Seaborg's office, in a connecting building. Seaborg suggested we go there to meet some of his small team of research students and postdoctoral fellows. As we walked across, Seaborg said that the elements beyond 106 were expected to decay, and rapidly so, chiefly by spontaneous fission—their nuclei splitting apart even without being hit by neutrons. But nuclear physicists had looked at the model of the atom—chemically inert when its electron shells are exactly filled—and by arcane calculations had come to a model of the nucleus itself also made up of shells, in this case shells of protons and shells of neutrons.

Upstairs, Seaborg introduced two of his research students and a postdoctoral fellow—David Morrissey, who was plump, open-faced, and relaxed, and who identified himself as a chemist turning into a physicist. "The way things go here, you have to think out experiments very carefully beforehand," Morrissey said. "And if it involves chemistry, you'd better get your chemistry going, well before your experiment actually comes up—because you might only get one or two cracks at the accelerator. You want to be confident and set to go, because you probably would not get the time to repeat an experiment."

Then Seaborg went on. "The periodic table is based on atomic structure and periodic variations, but in order for an element to be detectable either by synthesis or in nature it has to have nuclear properties that are suitable," he said. "It must be resistant to radioactive decay. Now, the noble gases, helium and argon and krypton and so on, are chemically nonreactive, which is another way of saying that they have extra stability because they have closed shells of electrons. In the same sense, there are closed shells of neutrons and there are closed shells of protons, in the nucleus, well known in the periodic table." These made elements that are unusually stable. Indeed, some elements had filled proton shells and filled neutron shells as well, and were superstable. Lead with 82 protons and 126 neutrons, atomic weight 208, is one of these. "What happened here was that about twelve years ago theorists predicted that there should be such a closed shell when there are 114 protons," Seaborg said. "And another one when there are 184 neutrons. And, indeed, you might have a doubly closed shell there at that nucleus that has 114 protons *and* 184 neutrons, which, of course, would have a mass of 298. And the prediction is that there should be an 'island of stability' around those magic numbers. So that would be element 114 with a doubly magic nucleus. But the 'island of stability' ought to stretch from 114 to 110, or something like that."

How confident were they that more elements could be made?

"That varies from person to person," Morrissey said. "Some persons, Dr. Seaborg among them, are very, very confident. This work is based solely on the extrapolations of nuclear theorists. They took the known properties of the known nuclei and made the logical extension of that knowledge and said that there should be a group of elements, right around here, that we should be able to make and that should live long enough that we should be able to see them in the laboratory. Then a number of chemists made their predictions about what the elements' chemistry would be, based on their electron configurations—their position in the periodic table."

"Element 114 will be like lead," Seaborg said. "It should be eka-lead. And element 112 should be eka-mercury, element 110 should be eka-platinum, and so on. And then there will be still another set of rare earths, the superactinides, starting at 122. But their nuclear properties are not expected to make them long-lived enough for identification to be possible."

"But predicting the chemistry is made complicated by the size of the atoms themselves," Morrissey said. "Because the nuclei are so large, we get some special effects of relativity." Predictions from a different theory intervene and affect the model. "Electrons have to travel around the nucleus with a velocity that depends on how far out they are. If you get the nucleus very large, and with a large number of positive charges—protons—and therefore a large number of electrons circling, the point is reached where the electrons are moving so fast that they begin to approach the speed of light. And Einstein's relativity says that then the energy of the electrons changes. This affects the filling of the electron shells—and if you are affecting the filling of the electron shells, then you are automatically affecting the chemistry of the elements. What happens is that the elements become more noble—more like the noble gases or the noble metals, gold or platinum, for example. So if you just count down the table to 114 you should get something that behaves like lead; but then you have that second, subtle prediction—from relativity—that you have to take into account."

I said that I had seen a report that some of the superheavies might have occurred naturally—that traces of their past presence had been found.

"Oh, in Madagascar!" Morrissey said. "The group was

at Florida State University in Tallahassee. People had been looking for superheavy elements, and here these fellows said that they had found what they reported was the evidence of the decay, long ago, of a superheavy element, in a sample of mica from Madagascar. And when that work first came out, and was published, it generated tremendous excitement in the scientific community. The case was not very strong—but it was a good case, and everybody was saying, 'How did we miss it?' But as time went on their case began to unravel bit by bit, though the work has never been retracted or demonstrated to be incorrect.

"It looked like evidence, at the start," Morrissey went on. "One sees a piece of work published in a reputable journal and the authors are reputable people—and you take it at face value. And then other people say, 'Let's see if we can reproduce that.' Or, 'If their work is true, we ought to be able to see this sort and that sort of extension of it.' And the work goes up, slowly, as the evidence mounts, or it goes down as the evidence decreases. And that case was one in which the work just slowly sank. It's part of the feedback process."

"They surely made a mistake," Seaborg said. "There are, however, claims from the Soviet Union, by Flerov and his group at the Dubna Laboratory, that they have found superheavy elements, one or more, on the evidence of spontaneous fission activity. They say they found them in meteorites, like the Allende meteorite that fell in Mexico about ten years ago, and they report the presence of superheavy elements also in certain kinds of ground water. Isotopes of elements in the island of stability around 114. But I doubt that the results Flerov reported are correct. To be detectable in meteorites or ground water—naturally occurring—the element would have to have a half-life of a hundred million years or more. Other calculations suggest half-lives quite a bit shorter—maybe as short as a year, or even less. Maybe only minutes or seconds."

Were there further islands of stability, yet farther out?

"There might be," Seaborg said. "There are indications that there might be one way up at atomic number 164—but there's no conceivable way to synthesize such elements, so that is not a very practical prediction."

Glenn Seaborg

8

Evidence

"It is also a good rule not to put too much confidence in the observational results that are put forward *until they are confirmed by theory,*" Sir Arthur Eddington wrote in 1934. Science does not deal in facts, as such. It demands significant facts. What makes a fact significant is its place in a developing pattern—or its refusal to take a place, its insistence that a new pattern be constructed that will include it comfortably. Indeed, if a theory is alive and progressing, scientists deliberately ignore evidence that doesn't seem to fit. Science deals with the interplay of facts with ideas. Eddington also wrote, in that same passage, "Observation and theory get on best when they are mixed together, both helping one another in the pursuit."

Arnold Barry Moskowitz is a convicted felon. On Thursday, June 3, 1976, a few minutes before nine o'clock in the morning, Michael Gleason, manager of a branch of the Chemical Bank of New York in the Sheepshead Bay section of Brooklyn, drew up his car by the back entrance to the bank building and got out. Nearby, a red Ford was parked, and in it sat a young man with a mustache. The young man called out, "Hi, Michael." Gleason looked over, but did not recognize the man. The young man got out of the car and showed Gleason a .38 pistol, and said something like, "Do you recognize this?" The man was also carrying a hand grenade and a walkie-talkie, and told Gleason that he and an accomplice were monitoring police calls. At gunpoint, the man took Gleason into the bank, and then ordered him and the assistant manager, Joyce Pyle, to open the vault. In the vault were two large canvas sacks awaiting delivery to the Federal Reserve system that morning. The young man seized them and left. The entire robbery took less than ten minutes. In the two sacks was $279,000—making this the biggest bank robbery in New York in the decade.

Police described the gunman as blond, with a

mustache, about thirty-five years old, 5 feet 8 inches tall, weighing 160 pounds, and wearing a brown jacket. On the day of the robbery, FBI agents took descriptions of the robber from Gleason and Pyle and from other bank employees who had caught a glimpse of him. The descriptions conflicted somewhat; they agreed, though, that the robber had blond or dark-blond hair and a blond mustache. His height was remembered differently, and none said that he was taller than 5 feet 10 inches. None of the bank employees mentioned the robber's complexion. Next day, from the descriptions, a police artist drew a composite sketch of the robber. One other fact was apparent: the robber had a good knowledge of the workings of the branch. For example, he had stayed near the vault, avoiding the main part of the banking floor, which was surveyed by videotape cameras; he knew Gleason's home address; and he appeared on a day when an unusually large amount of cash was packed for pickup.

Police sketch in hand, a security officer went through previous videotapes of the bank's customers. He came across a picture of one customer that three bank employees thought looked like the robber. By comparing the day and time of that picture with bank records, the customer was identified as Arnold Moskowitz. Moskowitz was twenty-one years old and a senior at Brooklyn College; he was living with his parents three blocks away from the branch, was a frequent customer there, had no police record, and was about to start a job as a salaried campaign worker in the reelection campaign of the then Senator James Buckley of New York. Moskowitz was 6 feet tall, weighed about 160 pounds, had brown hair and a medium-thick brown mustache—and conspicuous acne.

The FBI put Moskowitz under surveillance, then arrested him on August 12. They found no incriminating evidence in a search of his room at his parents' house; no trace of the money was found, no gun, no walkie-talkie, no red Ford; no accomplice was named. A police lineup was arranged, putting Moskowitz with five other men with mustaches. Three bank employees who had been with the robber in the vault, eleven weeks earlier, and two others who had gotten glimpses were brought to the lineup. Two of the employees picked Moskowitz out of the lineup. Two others picked an FBI man. The fifth bank employee said that none of the men in the lineup was the robber. Mainly on the evidence of the two positive identifications, Moskowitz was arraigned for bank robbery. He was released on $100,000 bail, put up by his parents with their house as security.

Moskowitz came to trial in July 1977, more than a year after the crime. He pleaded not guilty. Gleason, the branch manager, was the strongest prosecution witness. In the courtroom, he dramatically pointed to Moskowitz as the gunman. Joyce Pyle, the assistant manager, said Moskowitz was the robber—but admitted, under cross-examination, that at the police lineup she had picked the FBI man, and that she had originally described the robber's eyes as blue, though Moskowitz has brown eyes. Frances Criscione, a clerk at the bank, also identified Moskowitz—but conceded that her description of the robber the day of the crime did not closely resemble Moskowitz. For the defense, another bank employee, Kenneth McGuinness, said that he had seen the gunman's face for at least three seconds, that he had once previously met Moskowitz as a customer, and that the two were not the same man.

Pause for a moment: in you, the reader, the very telling of this story in this manner has, almost certainly, set up an expectation—a pattern of the facts, a theory about the case. Almost certainly, your theory is that Moskowitz was mistakenly identified. You may well be right. But the jury thought differently. Moskowitz was convicted of the crime. There was no evidence that unequivocally proved his innocence, either—and the jury had not even thought there was reasonable doubt. Moskowitz was sentenced to a prison term. He appealed. The appeals court, in May of 1979, reduced his sentence to three years on probation.

For many years, the phrase "circumstantial evidence" has had the stink of uncertainty about it. A prosecution case that's said to be merely circumstantial is thought to be provisional, not solidly proved, not watertight, no matter how closely woven the net of circumstance in which the defendant appears caught. In contrast, direct and singular eyewitness identification has been held by the law to be essential to conviction in the gravest kinds of crimes, has been treated by judges as admissible without serious question, and has been felt by juries—especially when the defendant is identified as the criminal by a witness who was a victim of the crime—to be overwhelmingly persuasive, outweighing almost anything else including strong alibis. In fact, however, of all the kinds of evidence heard in criminal trials, eyewitness testimony and especially eyewitness identifications are by a long way the least reliable.

Proof of the unreliability of the direct evidence of the eyewitness has been mounting, and from two sides. Some scholars of the law have long been troubled by the problem of mistaken identifications. In 1932, Edwin Montefiore Borchard, who was a professor at Yale University Law School, reviewed the cases and trials of sixty-five people who had been convicted of crimes but who were later proved innocent. "Perhaps the major source of these tragic errors is an identification of the accused by the victim of a crime of violence," Borchard wrote; and he went on:

> This mistake was practically alone responsible for twenty-nine of these convictions. Juries seem disposed more readily to credit the veracity and reliability of the victims of an outrage than any amount of contrary evidence by or on behalf of the accused, whether by way of alibi, character

> witnesses, or other testimony. These cases illustrate the fact that the emotional balance of the victim or eyewitness is so disturbed by his extraordinary experience that his powers of perception become distorted and his identification is frequently most untrustworthy. Into the identification enter other motives, not necessarily stimulated originally by the accused personally—the desire to requite a crime, to exact vengeance upon the person believed guilty, to find a scapegoat, to support, consciously or unconsciously, an identification already made by another. Thus doubts are resolved against the accused. How valueless are these identifications by the victim of a crime is indicated by the fact that in eight of these cases the wrongfully accused person and the really guilty criminal bore not the slightest resemblance to each other, whereas in twelve other cases, the resemblance, while fair, was still not at all close.

Many legal studies since have documented the problem, though few with Borchard's eloquence. Recently, social psychologists have been studying eyewitness evidence, both in the laboratory, to find out why it so often goes wrong, and in the courtroom, to find out what juries make of it. The conclusions are, if anything, more disturbing than those of the lawyers. Elizabeth Loftus is a psychologist at the University of Washington, in Seattle, who has become one of the leading authorities on the problems of eyewitness testimony and its evaluation in court; she has often appeared as an expert witness on behalf of defendants in criminal cases—though judges, indeed, are often reluctant to agree that expert testimony is necessary—to make clear to juries the dangers of eyewitness evidence. Loftus has recently published a book about the problems, titled *Eyewitness Testimony.* She tells of scores of cases, each more bizarre than the last, in which juries convicted on eyewitness identifications—even when made at fifty yards on a dark and rainy night, with no more light than the flare of a match, by a witness who turned out to be nearly blind. She moves to the stages before the courtroom, making palpable the great and almost unavoidable pressures on potential witnesses to simplify what they actually saw, to come up with some sort of identification even if they are not certain, and then to stick firmly to the identification. Loftus writes, in conclusion:

> The problem is clear: The unreliability of eyewitness-identification evidence poses one of the most serious problems in the administration of criminal justice. . . . The number of mistaken identifications leading to wrongful conviction, combined with the fact that eyewitness testimony is accepted too unquestioningly by juries, presents a problem for the legal community. . . . Excluding unreliable identifications, requiring corroborating evidence, and issuing cautionary jury instructions are three partial solutions which fail to provide an adequate answer.

The brute fact, the single, direct, positive, apparently decisive piece of evidence, thus turns out to be—even in the court of law—the least reliable basis for judgment. What's required is a complete and circumstantial pattern in which the eyewitness testimony has standing if it fits. In effect, the net of circumstantial evidence is the theory of the case.

The relation of the singular piece of evidence to the pattern is complex, shifting, many-shaded. It is full of ironies. Sometimes a piece of evidence must wait years, even decades, before theory finds a place for it. One remarkable astronomical sighting had to be rediscovered 798 years later for its standing to be decided. In the year 1178, the monk Gervase of Canterbury, who was keeping the chronicles of the cathedral and monastery, recorded an astonishing observation. "Hoc anno, die Dominica ante Nativitatem Sancti Johannes Baptistae," Gervase began, in his muscular, understated medieval Latin: "This year, on the Sunday before the birthday of St. John the

Baptist"—therefore on June 18—"after sunset when the moon was first visible a marvellous sign appeared to some five or more men who were sitting there facing the moon." Gervase went on:

> For the new moon was bright, and, in the usual manner of new moons, extending its horns to the east; and behold! suddenly the upper horn split in two. And from the middle of this division a flaming torch sprang up, casting fire, coals, and sparks a long distance. Meanwhile, the body of the moon, lower, writhed as though anxiously, and, to use the words of those who told me about it and who saw it with their own eyes, the moon wriggled like a snake that has been struck. After this, it returned to its proper state. This transformation repeated a dozen times or more, the flame tormented into different shapes, and finally returned to normal. After these transformations, the moon from horn to horn—that is, for its whole length—became blackish. And as I write this I recall those men, who saw it with their own eyes and are prepared to swear on their good faith and judgment that in what I have written above nothing false has been added.

A naïve account of a terrifying omen? The effects of the strong ale for which Kent was famous? Gervase made nothing more of the story. His moon, sun, planets were Aristotelian and moved in crystalline spheres unpierced and eternally unchanged. But in 1976 Jack Hartung, who is a planetary scientist at the State University of New York at Stony Brook, Long Island, was browsing in a recent monograph that catalogued observations of eclipses in medieval chronicles, and came across a fragment of Gervase's description. Hartung's specialty is impact cratering. He surmised at once that he was reading an eyewitness account of a lunar eruption or of the impact of an enormous meteorite on the moon. He sought out the full original and got it translated. A meteorite was more consistent with the present knowledge that the interior of the moon has been cold for many millions of years, and with studies of impacts in the laboratory and of large impact craters in the earth. Thus, when the sunlit crescent had seemed to split in two, part of it was hidden by the spewing cone of matter thrown up by the impact. The fire and sparks were incandescent gases and other matter released and thrown out. The wriggling was caused by gases in a temporary cloud of varying density, refracting moonlight like heat shimmer on a country road.

In the computer, it was easy to spin the moon back in time in order to be precisely sure what part was visible on that day at that hour and place. The moon that evening had been just forty hours past new, an extreme crescent. Working back from the crude limits of size and direction of the impact described, Hartung found one excellent candidate. The search required photographs taken from Soviet and American lunar spacecrafts—for the crater is over the edge, on the far side of the moon. The crater is very large, 20 kilometers across, and in a plausible position. It was first spotted in 1959, by Lunik 3; in 1961 it was named Giordano Bruno—after the mystical monk who was burned at the stake in Rome in 1600 for heresies that included support of the Copernican, heliocentric theory of the solar system. The crater Giordano Bruno and the bright, prominent pattern of rays extending far out from its walls were superimposed upon other, older craters and ray systems and were themselves pristine, which had independently led selenographers to think that the crater had been made more recently than other lunar features.

Hartung published his analysis, with some excitement—one rarely sees exclamation points in scientific journals—in September of 1976. It drew opposition, chiefly from those who thought the passage of a large meteor between earth and moon was a more probable event. It drew support, as well. Odile Calame and J. Derral Mulholland asked two questions. (She was at the Center for Research in Geodynamics and Astronomy, in Grasse, in the South of France, and he was visiting there from the McDonald Observatory at the University of Texas.) First, would the impact that formed the crater Giordano Bruno have caused effects that would have been visible, as the medieval eyewitnesses reported? They figured the size of the meteorite, the mass of the gouts of rock and other matter thrown out, their trajectories and distances in the lunar gravity—careful calculations, fundamentally simple, and yielding an unequivocal *yes*. The event would have been visible.

Second, more subtly, would such a collision have disturbed the moon's motion in ways still observable today? The moon would have rung like a struck bell—but such vibrations damp down quickly. The moon would also have picked up wobbles, oscillations in its rotation called free librations—and these have damping times much longer than eight hundred years. In effect, Calame and Mulholland realized that Hartung's hypothesis imposed a prediction, namely, that a component of the free librations should be found that could be traced to the Bruno collision dated eight hundred years ago. These motions are tiny; but the astronauts of the Apollo 11 mission who landed on the moon left behind instruments that have permitted, ever since, a series of measurements of the moon's motion by laser beams, and Mulholland, in Texas, had carried these out and analyzed them. The amount of oscillation measured .0005 degree—and was, in fact, an inexplicable anomaly, until Hartung's idea. In a paper in *Science* in February of 1978, Calame and Mulholland said, flatly, "The results of the laser analyses are only explicable by a recent impact. . . . What we have done is to show that such an impact [as supposed by Hartung] would have been observable and that the only modern observations that are capable of revealing the dynamical vestiges of such an event provide a compatible result."

The honesty and sobriety of five unknown Englishmen and a careful, credulous monk have been vindicated after eight centuries—and that's a sentimental comedy with a serious moral. Made surprising only by the unusual span of time, the sequence from Gervase through Hartung to Mulholland is a small jewel of the scientific process. A rogue observation, if it attains respectability at all, must wait for a context of theoretical ideas, both local and more general, and of other observations that can convert it from anomaly to natural consequence—from *that can't be!* to *of course that follows!*

Indeed, a sharp distinction can rarely be drawn between evidence and theory. "Observation and theory get on best when mixed together," Eddington said—but in living science they can hardly be separated. Observation is always flavored and sometimes saturated with theory. Galileo made his opponents look foolish. When he announced that through his early telescope he had seen mountains on the moon, spots on the sun, and four moons circling Jupiter, the public was astounded and fascinated. When he called Jupiter's moons the Medicean stars, in honor of the Grand Duke of Florence who he hoped would be his patron, a letter arrived from the court of the King of France, bidding Galileo, "Pray discover as soon as possible some heavenly body to which his Majesty's name may be fitly attached." When he claimed that his evidence confounded Aristotelian ideas and supported the Copernican view of the universe, the public listened with interest, and churchmen and other Aristotelians were alarmed. Several of them refused to concede that Jupiter had moons. Galileo invited them to look through his telescope—and ever since, these men have been figures of ridicule because they refused to behold the evidence for themselves. When one of them died—Giulio Libri, of the University of Pisa—Galileo mocked, "Libri did not choose to see my celestial trifles while he was on earth; perhaps he will now he has gone to heaven."

The polemic cleverly displaced the argument, for

Galileo's contemporaries had correctly seen his weak point. They were doubting not the evidence directly but the theory on which the evidence relied—the *optical* theory. They said that the claim that Jupiter had moons depended entirely on the reliability of Galileo's telescope, which they violently attacked, and on the optical theory of the telescope, which in that day was, in fact, almost nonexistent. Imre Lakatos, a Hungarian logician and philosopher of science who taught at the London School of Economics from 1960 until his premature death in 1974, wrote, ten years ago, "It was not Galileo's—pure, untheoretical—*observations* that confronted Aristotelian *theory* but rather Galileo's 'observations' in the light of his optical theory that confronted the Aristotelians' 'observations' in the light of their theory of the heavens." Lakatos pressed on to the general, powerful point about the nature of evidence: "There are and can be no sensations unimpregnated by expectation and therefore *there is no natural (i.e. psychological) demarcation between observational and theoretical propositions.*"

Galileo won the argument in the long run, of course—but scientists with a winning theoretical program are, if anything, even more bold than losers in ignoring evidence and preferring theory. Francis Crick, the leading theoretician in molecular biology, once explained to me the reason for that; sitting in his office in Cambridge, England, we had been talking about his discovery, with James Watson in 1953, of the structure of DNA. "You must remember, we were trying to solve it with the fewest possible assumptions," Crick said. "There's a perfectly sound reason—it isn't just a matter of aesthetics or because we thought it was a nice game—why you should use the *minimum* of experimental data." He mentioned an attempt by some colleagues to get a structure for proteins, which had gone embarrassingly wrong. "We knew that they had been *misled* by the experimental data. And therefore every bit of experimental evidence *we*"—he and Watson—"had got at any time we were prepared to throw *away*, because we said it may be misleading. . . ." Thinking out loud, excited by the point, Crick got up and began to pace back and forth, with long, loping steps, speaking in the rhythm of his stride. "They missed the alpha helix because of that! You see. And the fact that they didn't put the peptide bond in right.* The point is that evidence can be unreliable, and therefore you should use as little of it as you can. And when we confront problems *today*, we're in exactly the same situation. We have three or four bits of data, we don't know which one is reliable, so we say, now, if we discard that one and assume it's wrong—even though we have no evidence that it's wrong—then we can look at the *rest* of the data and see if we can make sense of *that*. And that's what we do *all the time*. I mean, people don't realize that not only can data be wrong, in science, it can be *misleading*. There isn't such a thing as a hard fact when you're trying to discover something. It's only afterwards that the facts become hard."

In present-day theoretical physics, of all science the most abstruse, the relation between evidence and theory is the same. Recall what Murray Gell-Mann said: "You know, frequently a theorist will even *throw out* a lot of the data on the grounds that if they don't fit an elegant scheme, they're wrong. That's happened to me many times. The theory of the weak interaction: there were *nine* experiments that contradicted it—all wrong. Every one. When you have something simple that agrees with all the rest of physics and really seems to explain what's going on, a few experimental data against it are no objection whatever. Almost certain to be wrong."

Under the compulsion of theory, scientists will

---

*See chapter 6, "Modeling."

reach deep to find evidence. One spring day, seven years after the discovery of the structure of DNA, Crick and two other molecular biologists, his close collaborator Sydney Brenner at Cambridge and François Jacob visiting from Paris, were discussing a problem that seemed intractable: how the genetic information carried on the DNA is read out so that it can direct the specific construction of protein molecules and the rest of the organism. In one of the most remarkable cases of collective inspiration in science, in an hour or so of intense discussion of many strands of evidence they came up with the idea that there must be an intermediary molecule, a "tape" that would copy the information off the DNA and carry it out to the places in the cell where proteins were assembled. They later called this intermediary the "messenger." Jacob and Brenner, later that spring, went to the laboratory of Matthew Meselson, at the California Institute of Technology, where in a strenuous month of experimentation they demonstrated that cells do, in fact, contain something that corresponds to some of the predicted characteristics of the messenger. Meanwhile at Harvard, in hot competition, James Watson and a colleague, Walter Gilbert, with several others in Watson's laboratory, were also trying to isolate the messenger. Ten years later, Gilbert described to me the terrifying uncertainties that experimental work may entail.

"The major problem really was that when you're doing experiments in a domain that you do not understand at all, you have no guidance to what the experiment should even look like," Gilbert said. We were in his office, at Harvard, on a cold gray afternoon in February. He put his feet on a chair, and went on, "Experiments come in a number of categories. There are experiments which you can describe entirely, formulate completely so that an answer must emerge; the experiment will show you A or B; both forms of the result will be meaningful; and you understand the world well enough so there are only those two outcomes. Now, there's a large other class of experiments that do not have that property. In which you do not understand the world well enough to be able to restrict the answers to the experiment. So you do the experiment, and you stare at it and you say, Now, does it mean anything, or can it suggest something which I might be able to amplify in further experiment? What is the world really going to look like?

"The messenger experiments had that property," Gilbert said. "We did not know what it should look like. And therefore as you do the experiments, you do not know what in the result is your artifact, and what is the phenomenon. There can be a long period in which there is experimentation that circles the topic. But finally you learn how to do experiments in a certain way: you discover ways of doing them that are reproducible, or at least—" He hesitated. "That's actually a bad way of saying it—bad criterion, if it's just reproducibility, because you can reproduce artifacts very, very well. There's a larger domain of experiments where the phenomena have to be reproducible and have to be interconnected. Over a large range of variation of parameters, so you believe you understand something. And this was what was happening with the messenger experiment: one learned how to get something out of the cell, how to isolate something, how to avoid— In order to isolate that something, for example, you had to prevent its being broken down by enzymes. And even to formulate that, you have to see it by accident and then discover what you have to do in order to see it reproducibly. So there's a whole domain of creating the phenomenon."

Coaxing the evidence into view, creating the phenomenon—yet artifacts are very reproducible. The dangers are obvious. Even more than in law, in science the evidence of the observer is subject to powerful and subtle distorting pressures. A friend of mine who is a crystallographer remarked to me once that he was just as glad that he had made a certain crucial set of measurements several weeks before he started reading through published papers and found the clue to the fundamental physiological mechanism that the measurements revealed. "Otherwise . . ." He let the sentence hang. The discovery capped a lifetime's work. The measurements taken without knowledge of what they signified were free of the suspicion of unconscious bias in observing them.

The results of such bias are called "observer effects." They have been detected in cases where there can be no hint of deliberate deception. They occur in the exact sciences, like physics, as well as in the soft sciences, like social psychology. And they afflict the most eminent of scientists. When the light of the sun is spread out by a prism into a spectrum, the bands of color are striped also by some narrow, dark lines, which we now understand to mark the specific wavelengths at which light is absorbed by various elements in the sun—iron, magnesium, calcium, and others. Isaac Newton experimented with prisms and lenses in the years 1668–72, and based an entire theory of the nature of color and of light on the solar spectrum. He did not report the absorption lines. Was his equipment too primitive to reveal them? In 1961, William Bisson at M.I.T. built an apparatus like Newton's, reconstructed the original experiments, and showed that the absorption lines would have been plainly visible. Bisson and a colleague reported this in a note in *Science* in 1962, concluding, " 'We saw the lines,' and wonder why Sir Isaac Newton failed to achieve the distinction of being the founder of the science of spectroanalysis." The question provoked Edwin Boring to argue, later that year, that Newton's theory, not his apparatus, had no place for the lines; Newton's theoretical expectations blinded him to the evidence. Boring added, "To the observing scientist, hypothesis is both friend and enemy."

Even the simplest steps of selection of material can introduce bias. A plant geneticist found, on analyzing his own work and that of his assistants, that they were all unconsciously picking the best-looking plants first for study. Experts in the design of experiments conclude that when a random selection is needed—a random selection of anything, from soil samples to celestial photographs to telephone subscribers for public-opinion polls—then even the best-trained experimenters will introduce nonrandom biases and must learn to do their picking with such aids as tables of random numbers.

Counting, so basic and apparently so straightforward, is full of traps. Every ten years, as required by the Constitution, the United States counts its population—and in this century the Census Bureau has learned that certain groups are notoriously, systematically counted short. The short count creates serious problems. For example, large cities qualify for many kinds of federal aid that are scaled to their populations—but in the city centers, poor people, and especially poor blacks, tend to be less settled than others and therefore harder to find; fathers may be missing from families, unemployed and unmarried young adults easily slip through. Another group that makes large demands on social services but that is very hard to count, for obvious reasons, is illegal immigrants. Census totals always show, also, slight bulges for people whose ages are exact multiples of five years—rounded off to the nearest five or zero. Rounding off is a problem everywhere that data are recorded.

Even when the things counted are not trying to dodge, counting may be tainted by observer effects. A routine but important measurement in medicine and research is the blood count—the counting, under the microscope, of the number of blood cells, red or white, in blood samples of standard size. For many years, textbooks that told how to make the counts and interpret them also maintained that if techniques were being followed correctly, two or more samples from the same blood should not vary in cell count beyond narrow limits, the "maximum allowable discrepancies." And in practice, lab technicians regularly reported counts that kept within the limits. But then three pathologists at the Mayo Clinic, in Minnesota, checked the standard procedures by a more cumbersome but more accurate technique, in which they made an enlarged photograph of the sample and then used a needle-sharp stylus to punch a hole through each blood cell in the picture; the stylus was connected electrically to a recorder which automatically tallied the punches. They counted many series of blood samples—and found the discrepancies within series to be greater, at least two-thirds of the time, than the supposed limits. "Differences considered as too large to be allowed will occur very frequently (66 to 85 percent of the time) if counting is made precisely and recorded faithfully." In other words, many observers for many years were making and recording observations that agreed with their expectations—but not with the realities.

The most extensive and amusing episode of observer error, amounting to a grotesque collective delusion that infected many of the best scientists of the day, is N rays. At the beginning of 1903, René Blondlot, who was a distinguished physicist at the University of Nancy, in France, was studying X rays, only recently discovered. Scientists had been unable to polarize X rays. Blondlot was trying to overcome that obstacle when, in the radiation from his X-ray tube, he discovered—as he announced to the Academy of Sciences on March 23—"an entirely new type of radiation." To honor the University of Nancy,

Blondlot called his discovery N rays. Over the next months, in a series of papers in the *Proceedings of the Academy of Sciences*, Blondlot reported his successive discoveries of the remarkable properties of N rays. They were a component of many kinds of radiation, including most sources of light and heat. When N rays hit any luminous object—say, a small flame—they increased its brightness; thus, their presence could be detected by the eye. "Some people appreciate the increase in the brightness of a small source of light due to N rays straight away and quite easily, while others find that the increase is hardly perceptible, and only after repeated attempts do they manage to observe it with any certainty," Blondlot wrote. He then found that the sun was a source of N rays. And they could be stored by certain substances, for example quartz, limestone, and even bricks, which would then reemit them. Blondlot demonstrated these effects for many scientists. He went on to construct a spectroscope for N rays with aluminum lenses and prism, and measured the wavelengths of N rays. By February of 1904, photographs of the effects of N rays were announced. Other scientists of good standing by now had thrown their labs into the research and were reporting still other aspects of the radiation. More than fifty papers were published. In December of 1903, Augustine Charpentier, who was professor of biophysics at Nancy, announced that N rays had physiological properties; they could, for instance, excite the nerves to emit similar radiation, thus making possible the tracing of nerve activity from the brain throughout the body by means of a simple phosphorescent screen. By early 1904, important diagnostic possibilities were developed using N rays and the phosphorescence they excited to trace diseases of the nervous system and in various organs of the body. The French Academy announced the award to Blondlot of a gold medal and a prize of 20,000 francs.

Most scientists in other countries were skeptical. In the fall of 1904, several physicists were discussing the problem after a meeting in England, and they deputized one of their number, the American experimental physicist and spectroscopist Robert Williams Wood, to go to Blondlot's laboratory. Blondlot had no English; Wood elected to carry on the meeting in German, so that Blondlot and his assistant would feel they could talk confidentially to each other in French. The demonstrations were performed for the most part in a darkened room. Taking advantage of this, Wood quietly and boldly interfered with the setup of a number of the experiments, once even pocketing the aluminum prism of the spectroscope, then putting it back before the lights went up. The results were unchanged. The next morning, Wood sent off a letter to *Nature* giving a full account of his findings.

N rays were nonexistent. Blondlot may have deluded himself; he may have begun in error and gone on in terror, like the embezzler driven further into theft and deception when his stock-market investments go sour; it's been suggested that the assistant in his laboratory confected much of the evidence. But the cautionary part of the tale is the number of other scientists who saw the effects of N rays and in perfect good faith performed experiments and published the results.

Observer effects shade into what are called experimenter effects—where the expectations of the scientist change not merely the observation of the result but the result itself. Experimenter effects are wild and subtle in psychology and the other behavioral sciences. The place where they are probably most marked, however, and certainly most dangerous is in testing the effectiveness of new drugs. Physicians have always known that some patients can be helped to feel much better by medicine that has no pharamacologically active ingredient—sugar pills, placebos (from the Latin for "I shall please"). These patients may have genuine diseases, not merely hysterical symptoms. Pain, for example, is perfectly real—subjective only in the pure and necessary meaning of the word that nobody else can feel your pain—and yet the intensity of pain will vary with context and circumstances, and in some people can be lessened by a placebo. Patients, of course, bring to their encounters with doctors the expectation that they will be helped. Just as importantly, doctors get their most fundamental satisfaction from the effective use of their power.

It's obvious that when a new drug is tested placebo effects must be ruled out. A control is necessary. Half the patients, picked at random, must get the new drug, the other half an injection of distilled water or a bottle of tablets that look just like the real thing but are inert. But at least as early as 1844, members of the Vienna Medical Society realized that for measuring the effectiveness of drugs such a blind test is not blind enough. The expectations of physicians—the experimenters—can be communicated to patients in all sorts of ways, so that the one dosed with the real thing, without being told so explicitly, will nonetheless acquire more confidence that he will be helped by the drug than does the patient who gets the dummy. Related to that experimenter effect is an observer effect, where, later on, the physician or nurse assessing signs of disease in the patient or recording what the patient reports of his symptoms will be influenced by knowing whether the patient "should" have got some benefit or not. Intricate techniques of double, triple, total blind experiment have been devised, whereby the actual matching of patients to drug or dummy—their therapeutic status—is established secretly by one chief investigator, for example by preparing coded bottles of pills. Nobody who has any direct dealings with the patients—to dispense medicine, to determine results, or even incidentally—has any knowledge of any patient's therapeutic status. And the results are evaluated by another chief investigator before the coded assignments are revealed.

Total-blind pharmacological testing has become standard since the Second World War. Studies of scores of such pharmacological tests involving thousands of patients have proved that the total-blind test is the most important technological innovation in modern pharmacological research. And it works for the curious reason that the placebo effect is strengthened: when the person giving the medicine to patients does not know that the pills are inert, he expects a better result—and gets it. The crucial consequence: placebo effect, the psychological component in the patient's getting better, is of equal strength for the drug and for the dummy.

Deliberate fraud is rare, and rarely talked about, but not unknown. The most notorious this century is Piltdown Man. On December 19, 1912, at the Geological Society of London, Charles Dawson and Arthur Smith Woodward announced that a layer of mammalian remains and stone and bone tools had been found in a gravel pit at Piltdown, in Sussex—including an Early Stone Age human skull and jawbone. Dawson was a Sussex solicitor and an enthusiastic amateur scientist. Smith Woodward was keeper of geology at the British Museum and prominent in paleontology. The discovery came at the end of a decade when paleolithic cave paintings, tools, and human bones were being found in great quantity in France and Germany; Piltdown Man was a marvelous intoxicant to English national pride. *Eoanthropus dawsoni* was striking: a cranium that was clearly human, a jawbone strongly apelike, made the creature the candidate for the missing link that popularizers of evolutionary theory were calling for. Enthusiasm was great. Dawson died unexpectedly in 1916; a year later, Smith Woodward announced that before his death Dawson had, in fact, found a second Piltdown skull. Smith Woodward lived to 1944, and defended Piltdown Man vigorously. Posthumously, in 1948, he published a popular book with the splendidly naïve title *The Earliest Englishman*—and it contained, among other unconscious delights, the observation, about a large bone implement that had been found in the gravel pit in 1914, that he and Dawson "were surprised to find that the damaged end had been shaped by man and looked rather like the end of a cricket bat." Smith Woodward is said to have been a humorless man.

Piltdown Man was anomalous from the start; in the years following the discovery, the anomalies got worse. In 1925, Raymond Dart and Robert Broom published the first discoveries of another kind of early man, the australopithecines from Africa, and these had a small and apelike cranium but jaws and teeth that had evolved toward the human—in fact, almost the diametric opposite of Piltdown. As the pattern of man's descent was pieced together, Piltdown Man moved from the main line to a sideline of his own. Then in 1953, at a dinner in London after an international congress of paleontologists, Joseph Sydney Weiner, an anthropologist from Oxford, had his attention drawn to the fact that nobody knew where, exactly, Dawson had found the second Piltdown skull and jaw. Pondering that, Weiner began to wonder whether the cranium and jaw, of the first find or of the second, really belonged to a single creature—or whether they had come together by chance, or deliberately. He initiated chemical tests of the bones, and proved that skull and jawbone, in each case, did not belong together, and that all the pieces were modern. When the investigation was widened, the rest of the finds in the gravel pit in Piltdown were easily shown to have been planted there. Piltdown Man was the Piltdown Hoax.

Who was the hoaxer? Charles Dawson was certainly implicated. Though questions arose about where he got some of the planted materials and the chemicals and skill to give them the appearance of age, for a quarter of a century after the exposure the case rested. Then, unexpectedly, a new suspect was named. He was William Sollas, who had been professor of geology at Oxford at the time of the original discovery and until his death in 1937. Sollas was accused by his successor in the chair, J. A. Douglas, who had died early in 1978—but who had left a tape recording, made ten years earlier, explaining what he knew. The tape was played publicly for the first time at a symposium of paleontologists in the fall of 1978. Douglas made two sorts of points. The first was motive: Sollas had despised Smith Woodward as a pretentious and ignorant fool, and had several times held him up to public scorn in scientific debate. The second was means: Sollas had access to the right sorts of bones to fabricate Piltdown Man, and Douglas remembered particular instances when Sollas had, for example, borrowed apes' teeth from an Oxford department, or had received a packet of one of the chemicals with which the Piltdown skull was stained. Douglas had kept quiet out of regard for his predecessor, but had been persuaded to make the tape by a neighbor of his where he was living in retirement. The story may not be complete even yet: other tape recordings are said to exist, with further denunciations, but remain sealed until the deaths of their makers.

The Piltdown hoax, in the forty years from its launching to its exposure, was a continuous source of confusion and controversy for paleontologists. Even while the evidence was accepted, however, its place in the picture of the origins of man became dubious. In effect, a large part of the community of science had tacitly agreed to get on with arguing out the patterns that seemed to explain the rest of the evidence, even if Piltdown Man refused to fit. The damage was contained. In its consequences, Piltdown Man is, after all, one of the greatest and most delicious intellectual practical jokes ever attempted. The fun remains.

Some frauds matter more. In 1961, Sir Cyril Burt, an educator and psychologist, published a paper in *The British Journal of Statistical Psychology* with the title "Intelligence and Social Mobility." Burt was universally respected, an elder statesman of educational research, preeminent in his field—and his field was the measurement of the genetic basis of human intelligence. In particular, he was famous for his studies of twins. By finding pairs of identical twins that had been separated soon after birth and raised apart, psychologists hope to be able to distinguish between those human characteristics that are in large part hereditary and those that seem to be mainly set by familial and other environmental influences on the growing child. Twins reared apart are not easy to find, and Burt had located and studied more cases by far than anyone else. He had also investigated sets of fathers and children in which the fathers belonged to various social and occupational classes.

The chief purpose of Burt's studies was to see to what extent intelligence—or, at least, performance on I.Q. tests—is inherited. In his paper of 1961, he concluded that the evidence was very strong indeed that intelligence is for the most part determined by heredity. He also concluded that there was a strong correlation of intelligence with social class, children of lower-class background being hereditarily stupider than those of the middle and upper classes. The paper of 1961 was the culmination of his lifetime's work. It had immense repercussions. Most important, certain American psychologists and educators picked up the argument and applied it, in reverse, to differences not only in social class but in race. American black children, in groups matched for age and other factors with groups of white children, have been shown in many studies to perform less well, on average, on I.Q. tests.

The explanations have inevitably been

problematic. The question was whether these differences had any significance, I.Q. tests being crude, and whether they reflected anything besides cultural differences, I.Q. tests being deeply dependent on the language and the expectations—basically white and middle class—of the questions themselves. Burt's work provided grounds for claiming that black children perform less well on I.Q. tests because they are hereditarily less intelligent. The controversy was politically explosive, even violent—one educational psychologist, Arthur Jensen, who argued for a close link between race and I.Q., had to be protected by bodyguards for a year on the campus of the University of California at Berkeley, where he worked. Its implications for educational policy, in a country that calls itself democratic, were deep and disturbing.

Burt died in 1971. Then in 1974 an American educational psychologist, Leon Kamin, charged that Burt had fabricated some of the data in the twin studies. In the fall of 1976, a journalist in London claimed that collaborators Burt had credited with gathering information did not, in fact, exist. Then others began to question the peculiar perfection of Burt's statistical results: they were too close to ideal distributions to seem real. In September of 1978, Donald Dorfman, professor of psychology at the University of Iowa—in a department that has for many years specialized in the measurement of intelligence, among other forms of testing—published a thorough and devastating analysis of Burt's famous paper. Dorfman concluded, "A detailed analysis of these data reveals, beyond reasonable doubt, that they were fabricated from a theoretical normal curve, from a genetic regression equation"—in other words, the conclusion was put into the data—"and from figures published 30 years before Burt completed his surveys."

Robert Rosenthal, a social psychologist at Harvard, wrote in 1966, "It is difficult to imagine a field of science in which each worker feared that another might at any time contaminate the common data pool. . . . Science has a way of being very harsh with those who break the faith"—he mentioned one scientist who was driven to suicide—"and very unforgiving. A clearly established fraud by a scientist is not, nor can it be, overlooked. There are no second chances. The sanctions are severe not only because the faith is great but also because detection is so difficult. There is virtually no way a fraud can be detected as such in the normal course of events."

Evidence:
The State of the Art

The frontiers of science—most of the time in most sciences—are at least as likely to be set by limitations of technique as by limitations of imagination. By tradition, of course, pure science—science "for its own sake"—is contrasted with engineering and technology, basic research contrasted with applied research. The distinction has sometimes been attacked; there's no doubt that it encourages a kind of snobbery; yet it is valid in certain contexts. In the mid-1970s, for example, the United States Congress declared a "War on Cancer" that most good biologists thought was entirely premature: not nearly enough was known about the fundamental processes of life in mammalian cells to permit rational, sensible choices of research targets in such a war. What was needed was the careful funding of further basic research. Yet, in practice, science however pure is incurably dependent on techniques, methods, instruments: a large part of science *is* technology.

Again and again, a new device or a new level of observation has set scientists' imaginations free, or given theory the jolt and the discipline of the unexpected. Conversely, often the most important exercise of a scientist's imagination is not what he devotes to theory-making but to the development of new methods and apparatus. Theoretical physics and astronomy are unusual, in this respect, only in that this division of effort is obvious: such is the size and intricacy of a new particle-accelerator or a satellite-borne radiotelescope that theoretician and experimentalist can no longer be the same person. A young biochemist put it flatly: "You get there any way you can"—and watching him in his laboratory was like watching the chef in a Paris bistro. In this most direct sense, science is an art. It follows, too, that new science is often done at the extreme limit of technique, or of the performance of the instruments. Columbus had a rougher crossing than Lindbergh, after all; in science today the pace of technical development is sometimes so great that what won a Nobel Prize for the man who did it laboriously and crudely just fifteen years ago—for the first time—may now be performed elegantly and easily by a graduate student as part of a doctoral dissertation.

Scientific instruments are beautiful objects. A seventeenth-century microscope, a medieval Moorish astrolabe, an early clock that shows the development of the escapement, an orrery, an ancient Chinese compass—such things have become collectors' items. The craftsmanship is instantly apparent; its full fascination lies deeper, though. Consider, say, the apparatus, so simple and inexpensive that any skilled machinist can duplicate it, with which Ernest Rutherford demonstrated the splitting of the atom: such an instrument is the physical representation of a particular process in the scientific imagination. It is an idea crystallized.

New instruments have produced revelations that touched off revolutions. The first microscopes found a universe of life in a drop of ditchwater. Two hundred fifty years later, the electron microscope has disclosed things within cells that must be understood as functioning physiological structures—but on a scale thousands of times smaller than anything known to anatomists before. Of all sciences, though, the history of astronomy illustrates most dramatically the importance of advances in instrumentation. Johannes Kepler calculated the elliptical orbits of the planets from the data of Tycho Brahe—data that Tycho had gathered by observation with the naked eye, aided by quadrants and other devices he built himself. But Kepler's calculations were relatively abstruse. The revolution in astronomy was publicized and forced home by the telescope, turned on the skies by Galileo. Often it's not the original inventor of an instrument who puts it to its decisive first uses, but rather a scientist a pace removed from the problems of the new tool's development, who can see it in a wider imaginative context. In 1609, Galileo was professor of mathematics at Padua, in the Venetian Republic, when he learned that someone in the Netherlands had combined two lenses in a tube to produce an instrument that made distant objects appear closer. He figured out the principle and built a telescope of his own. He proceeded to make an extraordinary series of discoveries. He saw a multitude of new stars, and comprehended the true nature of the Milky Way—not a smudge of diffuse light but "nothing else but a mass of innumerable stars planted together in clusters." He saw spots on the face of the sun and mountains and craters on the surface of the moon—blemishes on the perfection of celestial objects that outraged traditional philosophers but that made the heavens suddenly and for the first time comprehensible in down-to-earth terms. He discovered the four largest moons of Jupiter and recognized their revolutions as a miniature of the Copernican model by which the planets revolved around the sun. Later he found that Venus exhibits phases, like the earth's moon, from crescent to half to full—which it must do, by the Copernican model, since its orbit around the sun lies inside the earth's. He thus confirmed a strong prediction of the Copernican model. The rhetorical power of Galileo's discoveries—their power to convince other scientists and to interest and persuade the educated laymen of his day—cannot be understated. Three and a half centuries later, the invention of radiotelescopes has led to an upheaval in astronomical ideas nearly as thoroughgoing, and a public fascination nearly as great—except that, as yet, no model encompasses the observations in the far reaches of the universe as the Copernican model did for the solar system.

The crucial importance of technique in science is demonstrated by the frequency with which Nobel Prizes have gone to those who developed new apparatus or new methods. For eighty years now, Nobel Prizes have been awarded for discoveries in physics, in chemistry, and in physiology or medicine. Many of the prizes have gone for just the sorts of science one naturally thinks of as a great discovery. James Chadwick found the neutron, for which he got the prize in physics in 1935. Tsung Dao Lee and Chen Ning Yang conjectured and proved in 1956 that in certain subatomic interactions the universe does not behave with perfect symmetry; they overthrew the concept called "conservation of parity" in those interactions, a result so strong and fundamental that Yang and Lee got the prize in physics just one year later. But often a prize has been awarded for advances that were essentially technical—developments that led to discoveries of which the most important, as with Galileo and the telescope, were like as not made by other scientists. Repeatedly, such new techniques have touched off a cascade of Nobel Prizes. Once, to be sure, the judgment of the selectors was comically bizarre: in 1912, the Nobel Prize in physics was awarded to Nils Gustaf Dalén, a Swedish engineer, for his invention of automatic regulating devices for gas illumination in lighthouses!

9

Theory

In the dialogue between the possible and the actual, the search for solutions, what qualities of mind are required? Not long ago, I asked that of Joshua Lederberg, who is a geneticist and who got his share of a Nobel Prize for discovering the fact that some bacteria mate, passing copies of their DNA—their genes—from one to the other, and for following out the consequences of that discovery. We were in Lederberg's office at Stanford University; he was putting books and file folders into cardboard boxes, because in a few days he was moving to New York to take up the presidency of the Rockefeller University.

"I don't think there is one logic for science and another logic for the commonsense world," Lederberg said. "If there were, we would be in real trouble. I think there is a somewhat more systematic use of formal reasoning—well, it is more formal than everyday language and less formal than the mathematical type of reasoning. I'd say that the ability to discover analogies, the ability to generalize, the ability to strip to the essential attributes of some actor in the process—the ability to imagine oneself *inside* of a biological or other situation—these are some of the pretty obvious talents."

I mentioned the other sort of stripping away, that which creates the models of the mathematical physicist, the universe abstracted to a page or two of equations.

"Well, *we* are not ready to write equations, for the most part, and we still rely on mechanical and chemical or other physical models," Lederberg said.

What were such models about?

"When I think of a DNA molecule, I have a model of it looking like a rope," Lederberg said. "And I know, for example, if I pull at the two ends of a DNA molecule, it will break somewhere. Then I have to jump to a different level and say, 'Well, I know the structure isn't quite like that.' And I will try to look a bit more analytically at the question, 'Can I predict where the DNA molecule will break, if I hold the two ends, knowing that it isn't quite like a rope—in the following particular ways?'

"I think that that ability to move from one level of analogical reasoning to another, and not get stuck in the analogy at inappropriate times, is terribly important." Lederberg blew the dust off a book, put the book into a box. "You had better be able to do it. You have to be able to fantasize in rather crude ways—but then be able to shift from one frame of reference to another. That, I think, is more rare than you might suppose.

"Then there's a skill at combinatorial arrangements that comes up over and over again. Constantly, in planning experiments—trying to think if there are different ways in which a system might be put together—one has to have the skill to do a systematic, fairly rapid first scanning of the possibilities, of a given territory. And then the ability to know whether or not that's worth pursuing."

Was that in part a matter of trial and error?

"There are still enormous amounts of trial and error," Lederberg said. "Even after one has thunk through all the possible arrangements, one's still usually left with a few dozen—or a few hundred—testable alternatives. Which you then have to try and work through in a somewhat more systematic way. You may just come upon a blank wall. And then it helps, to make more discoveries, just to fiddle. You make more discoveries in the course of simply putting a couple of things together, in ways where you think you know how they are going to behave: but if the system is complicated, and you are not sure, then every now and then you get a discrepancy. Then you need the ability to spot a discrepancy quickly—that is, to know when you have got a finding that doesn't fit the model you thought you had. But many people are just not that explicit about what their premises were.

"That is one of the most important functions of experimental finesse. The person who is simply careless in experiments either ends up being a scoundrel, if he believes in his results and insists on them; or, much more usually, he is so mistrustful of the isolated finding that he will attribute it to an instrumental failing or to an error in technique. If you don't have the confidence that when you do an experiment you do it correctly, then when you get an aberration you're likely to say, 'Well, a speck of dust fell in,' and then ignore the result. Or else spend endless time reexamining what were, in fact, artifacts."

I said that there appeared to be a feedback between fact and idea—or rather, an upward spiraling between model and evidence.

"You go back and forth from observation to theory," Lederberg said. "You don't know what to look for without a theory; and you can't check the theory without looking at the fact; and the fact is only meaningful in the light of some theoretical construction."

Medawar had written of that back-and-forth movement of the mind, I said.

"I believe that that movement back and forth occurs thousands, even millions of times in the course of a single investigation," Lederberg said.

A model can be a kind of theory—even as simple a model as the sheet of paper that Linus Pauling folded and curled into his first three-dimensional approximation of the alpha helix in proteins, even as complex a model as the thousand-equation simulation of the United States' economic system that Lawrence Klein stores and runs in his computer. A model takes on the quality of theory when it abstracts from the raw data the facts that its inventor perceives to be fundamental and controlling, and puts these into relation with each other in ways that were not understood before—thereby generating predictions of surprising new facts. But models can serve the theorist in subtler ways, as stepping-stones to theories, as suggestive analogies, as tools with which to visualize ideas that would otherwise be too abstract to be seen clearly at first or stated confidently. Models or analogies of this sort are an intellectual framework, taken down and tossed away when the theoretical structure can stand on its own.

I suspect that no theories are erected without scaffolding of this kind. Certainly Einstein, for instance, began the line of thought that led to the special theory of relativity when, as a youth, he imagined himself riding a light wave. He began the first paper on the special theory of relativity with his crucial restatement of the meaning of simultaneity—in terms as limpid as any that Socrates ever proposed, and with the homely image of the train arriving at Zurich railway station at a particular time shown on his pocket watch. Throughout, the special theory is a matter of measuring rods that change size and clocks that run slow—and these are not the modern lecturer's patronizing visual aids, they were the instruments of Einstein's thought. Only in the general theory of relativity was his imagination less visual, more purely mathematical. The evolution from the special theory to the general theory took eleven years.

No other scientist has ever equaled Einstein's celebrity: the devotion that people of all sorts felt for his apparent wisdom, his sincerity, his simplicity, and the awe in which they held the supposed difficulty of his theories. And yet by physicists, while the depth, the power, the importance of Einstein's work are unquestioned, certain other intelligences of that extraordinary thirty years when modern physics was made—1900–1930—are generally considered to have been more austerely theoretical than Einstein. Among such figures, Paul A. M. Dirac is perhaps the one thought to dwell farthest up the mountain. And of all Dirac's ideas, his assertion in 1930 that the theory of relativity when applied to quantum mechanics required the existence of antiparticles—indeed, of what we now call antimatter—is supposed to have been most pristinely derived from formal, theoretical, mathematical requirements. In the spring of 1979, I asked Dirac about that, and got a surprising answer.

This introduction may make the theory sound terrifyingly hard: the proofs are a matter for specialists, to be sure, but the line of thought can be paraphrased simply, and nobody has done that better than Dirac himself in his Nobel lecture, given in Stockholm on December 12, 1933, under the title "Theory of Electrons and Positrons." Dirac was then thirty-one. He had been born in Bristol, of a Swiss father and an English mother; he had trained first as an electrical engineer, but had come to Cambridge to take a doctorate in mathematics, had stayed on, and was already Lucasian Professor of Mathematics—a post Newton had once held. His Nobel lecture, among other qualities, is short—and is amusingly arrogant about the standing of theory in science as perceived by a young man at the confident height of his powers.

Near the beginning Dirac said, "I should like here to discuss the simpler kinds of elementary particles and to consider *what can be inferred about them from purely theoretical arguments.*" This, no doubt, contributed to the growth of the legend of Dirac's approach. Anyway, he went on to explain how general physical principles—quantum mechanics with relativity put in—lead the mathematician to predict that there will be, in the real world, particles with exactly the mass, the spin, and the magnetic properties of the electron that was known to experimental physicists. Dirac said, in effect, that what is, had to be. "These results are in agreement with experiment. They were, in fact, first obtained from the experimental evidence provided by spectroscopy and afterwards confirmed by the theory." The words resemble Sir Arthur Eddington's about the interaction of evidence and theory, but the context and tone are different: Dirac was asserting without room for compromise the primacy of theory over evidence.

A moment later, Dirac mentioned another consequence of the theory, "a prediction which cannot be directly verified by experiment." Interesting here is not the detail of the unverifiable prediction itself, but Dirac's judgment: "One must believe in this consequence of the theory, since other consequences of the theory which are inseparably bound up with this one, such as the law of scattering of light by an electron, are confirmed by experiment." Thus, in passing, Dirac captured in a phrase what theories do: they compel belief because they bind diverse consequences together inseparably.

He then went on, "There is one feature of these equations which I should now like to discuss, a feature which led to the prediction of the positron." He pointed out that the equations for the electron allowed not one but two solutions: the electron could have positive kinetic energy—that is, its energy-in-motion—or negative kinetic energy. As far as the equations were concerned, the two were equivalent. (The analogy likely to be familiar is the square root of a positive number, which may be either positive or negative.) But electrons of negative kinetic energy were strange creatures indeed, Dirac said:

> In practice the kinetic energy of a particle is always positive. We thus see that our equations allow of two kinds of motion for an electron, only one of which corresponds to what we are familiar with. The other corresponds to electrons with a very peculiar motion such that the faster they move, the less energy they have, and one must put energy into them to bring them to rest.
>
> One would thus be inclined to introduce, as a new assumption of the theory, that only one of the two kinds of motion occurs in practice. But this gives rise to a difficulty, since we find from the theory that if we disturb the electron, we may cause a transition from a positive-energy state of motion to a negative-energy one, so that, even if

we suppose all the electrons in the world to be started off in positive-energy states, after a time some of them would be in negative-energy states.

Thus in allowing negative-energy states, the theory gives something which appears not to correspond to anything known experimentally, but which we cannot simply reject by a new assumption. We must find some meaning for these states.

The theory forced Dirac to suppose that "in the world as we know it" nearly but not quite all the possible places—technically, "states"—that can be occupied by electrons of negative kinetic energy are actually occupied, "and that a uniform filling of all the negative-energy states is completely unobservable to us." Elsewhere, he called this idea a "sea" of electrons of negative kinetic energy. Then came the leap of imagination: *"Any unoccupied negative-energy state, being a departure from uniformity, is observable."* Dirac went on, "An unoccupied negative-energy state, or *hole,* as we may call it for brevity, will have a positive energy, since it is a place where there is a shortage of negative energy. A hole is, in fact, just like an ordinary particle." A hole should be identical to the electron in every respect, except with a positive rather than a negative charge.

Dirac first put forward his theory of electrons in 1928, and his theory of holes in 1930. When he predicted the positively charged electrons, none had been noticed. Within two years, experimenters found the tracks of particles in their cloud chambers that behaved exactly like electrons except that in a magnetic field they curved in the wrong direction. They were electrons of positive charge—soon called positrons. By the time of Dirac's Nobel lecture, he was able to say:

From our theoretical picture, we should expect an ordinary electron, with positive energy, to be able to drop into a hole and fill up this hole, the energy being liberated in the form of electromagnetic radiation. This would mean a process in which an electron and a positron annihilate one another. The converse process, namely the creation of an electron and a positron from electromagnetic radiation, should also be able to take place. Such processes appear to have been found experimentally, and are at present being more closely investigated by experimenters.

In principle, there was no reason why the idea could not be extended to protons, which ought thus to have antiprotons. But because the proton is enormously heavier than the electron, a more complicated theory seemed likely to be necessary.

In any case I think it is probable that negative protons can exist, since as far as the theory is yet definite, there is a complete and perfect symmetry between positive and negative electric charge, and if this symmetry is really fundamental in nature, it must be possible to reverse the charge on any kind of particle. The negative protons would of course be much harder to produce experimentally, since a much larger energy would be required, corresponding to the larger mass.

If we accept the view of complete symmetry between positive and negative electric charge so far as concerns the fundamental laws of Nature, we must regard it rather as an accident that the Earth (and presumably the whole solar system), contains a preponderance of negative electrons and positive protons. It is quite possible that for some of the stars it is the other way about, these stars being built up mainly of positrons and negative protons. In fact, there may be half the stars of each kind. The two kinds of stars would both show exactly the same spectra, and there would be no way of distinguishing them by present astronomical methods.

These were astonishing ideas then; they still are. Dirac's prediction that for every kind of particle an antiparticle should be found has been amply borne out. The sorts of symmetries that he perceived have become for modern physics a guiding principle of almost magical power. And in the 1970s, with the rise of radio astronomy, galaxies were found that broadcast such stupendous amounts of energy that perhaps the only plausible explanation is that they are places where a vast, dense cloud of dust and gas and stars made up of our kind of matter is passing through a vast, dense cloud of dust, gas, and stars of antimatter, in mutual annihilation.

When I went to see Dirac in the spring of 1979, I reminded him of his insistence, forty-five years earlier, that "we must find some meaning for these states." That is often described as an idea in physics inspired by a purely mathematical and theoretical conception—and by no model, no experiment, by nothing but the abstract considerations. Was this actually the case?

"I think that would be true," Dirac said. He was then seventy-seven, a wisp of a man, precise, quick, and quiet, with a voice as dry as the rustle of an antique silk taffeta. He was still energetically pursuing theoretical physics. Several years before reaching the age of sixty-five and mandatory retirement at Cambridge, Dirac had accepted a lifetime job as professor of physics at Florida State University in Tallahassee; we talked in his office there. I asked whether he recalled exactly how he came to the idea of holes, of antimatter—besides the equations themselves.

Dirac paused, then spoke slowly. "I *was* thinking of the example of chemical valency," he said. Valency describes the willingness of a chemical element to combine with others, and was known to chemists before physicists explained it; the columns in the periodic table enclose elements of similar valency. "If you have an atom with all of its shells of electrons filled up, like one of the inert gases—helium, neon, and so on—it has no valency," Dirac went on. "Then you may have an atom with one electron more, outside the filled shells—with a valency of one. That applies to sodium, potassium, these alkali elements. Then there are atoms where you have a *hole* in one of the otherwise filled shells. They are chlorine, fluorine, bromine, and so on. They also have a valency of one. And it seems the hole in the electron shell *acts very much like a single particle.*

"I thought very much of that. And I think that that very strongly suggested to me that a hole in the negative-energy distribution of electrons would behave like a single particle. Which wasn't such a wild step! When you think of those chemical valency questions."

A theory, after the scaffolding of visualizations, analogies, and models has been taken down, is a structure of a special kind. To begin with, it is self-supporting, by the strength of the connections it makes in many directions—the diversity of the consequences that are, in Dirac's phrase, inseparably bound up with it. Furthermore, no theory stands alone. These structures are parts of larger sets of theories that mutually support and illuminate each other. There is, of course, an analogy for such a set of structures. The best model for theory is a map.

The map as metaphor for the network of scientific knowledge has often been suggested, but never with more wit and strength than in a recent discussion by the English physicist John Ziman.* Much more is meant than that scientific information and ideas are often best conveyed in diagrams. A theory, like a map, must fit together data that are inevitably incomplete, always at some level approximate, sure to contain mistakes. At the same time, the map connects up the information in many directions: the process of constructing a map tends to catch data that are dubious or strained, and the attempt to correct a map or to add new information reveals that it cannot be changed significantly at one place without forcing related adjustments in a wave spreading from nearby to far. "Analogously,

---

*See Ziman's fine book, *Reliable Knowledge* (Cambridge and New York: Cambridge University Press, 1978), especially pp. 82–85, to which I am indebted for several of the comparisons here—and even more for the pleasure of following the possibilities further.

scientific knowledge eventually becomes a *web* or *network* of laws, models, theoretical principles, formulae, hypotheses, interpretations, etc., which are so closely woven together that the whole assembly is much stronger than any single element," Ziman wrote; and he went on:

> This characteristic of any well-established body of scientific knowledge is often obscured by a conventional historical account of the initial phases of its discovery. Metaphorically speaking, the paths followed by the first explorers, passing by and recording many significant features of the landscape, are somewhat arbitrary, and may not cover the territory systematically. With only their notes to go by, we could imagine many alternative maps of the same region—just as some naïve critics of relativity or quantum theory suppose that their own homespun theories of space, time and matter need only be consistent with a few famous experiments . . . which happen to play an important part in the historical evolution of modern physics. This fallacy arises through ignorance of the innumerable detailed investigations that have since criss-crossed the territory and definitely resolved any major ambiguity in the original map—in this case, the whole body of research in atomic, nuclear and sub-nuclear physics of the past 50 years.

The metaphor of the map works so well because useful theories are not linear, like a cookbook recipe or a railroad timetable; the structures of science are interconnected in many dimensions. The multidimensional quality of scientific understanding has crucial effects. For one thing, it explains why it is so difficult to describe "scientific method" as though it could ever be reduced to a linear process; it explains in large part why scientists find most descriptions by historians or philosophers of science to be so peculiarly remote from their own experience.

At the same time, the connectedness is what guards science against error and fraud. An excellent instance is the flamboyant career, several years ago, of the magician Uri Geller, who claimed that by mental power he was able to bend spoons, bend keys, and stop clocks at a distance. The performances were amusingly contrived; the publicity was managed with brilliance. One wonders, as one always does looking back, at the silliness of the affair—bent spoons?—but at the time many people took Geller seriously. Even some scientists were credulous, while scientists who objected were challenged for their narrowness of mind by Geller's partisans and publicists—so that many kept quiet in order not to fuel the controversy and publicity. A rival magician, the Amazing Randi, demonstrated early and loudly that he could do everything Geller had done, using as his only mental power the simplest ruses of the conjurer's trade. Randi's exposure of the mechanics of the hoax was useful; what mattered far more—and yet was something difficult to get across to nonscientists—was that Geller's claims were unworthy of belief because, if true, not just isolated scientific notions but vast areas of the interconnecting map of the way things are would have to be abandoned. The simplest question—Where does the energy come from to bend the spoons?—reaches at once to the foundations of thermodynamics, the laws of conservation of energy and mass. In a more serious fraud, that of Piltdown Man, the faked skulls caused increasing strains and dislocations for many years, in conflict with all the accumulating evidence and theory about early man. Long before the hoax was uncovered, Piltdown Man was becoming provisional—was being moved to the margin of the paleontologists' mental maps.

Even in the making, a discovery emerges from the complex network of other evidence, theories, and models that the individual scientist possesses. Max Perutz recently reminded me that, more often than not, a discovery is made by the worker who possesses a combination of knowledge that nobody else can muster. "This is a further divide between mystical and scientific inspirations!" Perutz wrote.

From this, too, arises the extreme importance of collaborations in many scientific discoveries—and the importance of conversation, of travel, of visits by scientists to the laboratories of colleagues and competitors. Collaborations combine knowledge and skill in new ways. For just that reason, even the most intense collaborations in science rarely last more than a few years: as present problems are solved, new ones are reached that require different combinations. For similar reasons, great laboratories become great and remain so for a while because, among other reasons, they attract the right visitors.

Repeatedly, when one analyzes the fine grain of the events leading to some discovery, one sees what I call the "shuttle effect"—in which the physical arrival of a new person brings together the necessary elements. When James Watson came to the Cavendish Laboratory, in Cambridge, England, in the fall of 1951, he brought with him a detailed knowledge of genetics that nobody there possessed. He found there a level of skill at determining the structures of molecules by X-ray diffraction that was unequaled—and, in particular, he found Francis Crick, who was creating the mathematical tools for determining the structures of large helical molecules. The collision of genetics with X-ray analysis made possible the discovery of the double helix, of the molecular structure of DNA, the genetic material. In another case, the shuttle effect required a move of no more than fifteen yards. In the early 1950s, in an attic at the Institut Pasteur in Paris, Jacques Monod and François Jacob were working in laboratories at opposite ends of a short corridor. Though no one there realized it, the problems they were working on were intimately related. Then each problem began to help the solution of the other; a collaboration began that was one of the most intense and long-lived in modern science and that led to the working-out of how the information in the genes is read out and controlled. When that collaboration came to an end, after a decade, the breakup was as bitter and inevitable as a divorce.

Once the discovery is made, other people begin to search for other routes, from other starting points, to the same place: the looms of corroboration weave the discovery into the fabric of scientific knowledge. This process of connecting up the new seldom fails; it is, Ziman also wrote, "the source of the extraordinary reliability that scientific knowledge can often attain in practice."

Science is thus not only theories interconnected with each other and with the evidence, but the interconnections of the people who work with them. People and ideas taken together, the interconnectedness itself becomes the instrument of growth. For this reason, the logician Imre Lakatos said, in a radio broadcast in 1973, "I claim that the typical descriptive unit of great scientific achievements is not an isolated hypothesis but rather a research program." Lakatos went on:

> Science is not simply trial and error, a series of conjectures and refutations. "All swans are white" may be falsified by the discovery of one black swan. But such trivial trial and error does not rank as science. Newtonian science, for instance, is not simply a set of four conjectures—the three laws of mechanics and the law of gravitation. These four laws constitute only the "hard core" of the Newtonian program. But this hard core is tenaciously protected from refutation by a vast "protective belt" of auxilliary hypotheses. And, even more importantly, the research program also has a "heuristic," that is, a powerful problem-solving machinery, which, with the help of sophisticated mathematical techniques, digests anomalies and even turns them into positive evidence. For instance, if a planet does not move exactly as it should, the Newtonian scientist checks his conjectures concerning atmospheric refraction, concerning propagation of light in magnetic storms, and hundreds of other conjectures which are all part of the program. He may even invent a hitherto unknown planet and calculate its position, mass and velocity in order to explain the anomaly.

Lakatos was speaking, in part, about how one can tell the difference between science and pseudoscience. The test he proposed was the power of a research program to make those astonishing predictions. "In degenerating programs," he said, "theories are fabricated only in order to accommodate known facts." The most egregious instance of pseudoscience, Lakatos suggested, is Marxism. The predictions of Marxism have, indeed, been notoriously unsuccessful:

> It predicted the absolute impoverishment of the working class. It predicted that the first socialist revolution would take place in the industrially most developed society. It predicted that socialist societies would be free of revolutions. It predicted that there will be no conflict of interests between socialist countries. Thus the early predictions of Marxism were bold and stunning but they failed. Marxists explained all their failures: they explained the rising living standards of the working class by devising a theory of imperialism; they even explained why the first socialist revolution occurred in industrially backward Russia. They "explained" Berlin 1953, Budapest 1956, Prague 1968. They "explained" the Russian-Chinese conflict. But their auxilliary hypotheses were all cooked up after the event to protect Marxian theory from the facts. The Newtonian program led to novel facts; the Marxian lagged behind the facts and has been running fast to catch up with them.

A vital new idea, a living theory, by predicting astonishing new facts does one thing more. By doing that successfully, it draws able people to itself—as Hermann Minkowski and Sir Arthur Eddington, among scores of others, were attracted to the theory of relativity and then made crucial contributions. Thus the theory builds up around itself a program of research. And thus a living theory, precisely because it makes dramatic predictions of new evidence, generates new theory, too. This is the dynamism of scientific revolutions.

The interconnectedness of science tells us also that theories are not free speculations. They face rigorous restrictions. But more than that, they absolutely *require* restrictions. Dirac, in his Nobel lecture, said, "When one subjects quantum mechanics to relativistic requirements, one imposes restrictions on the properties of the particle." The restrictions were what did it: "In this way one can deduce information about the particles from purely theoretical considerations." But the need for restrictions—and the need to find the right ones—applies in the making of theory anywhere in science. One of the simplest but most dramatic examples was (once again) Pauling's understanding that in any structure for proteins the bond that joined each amino acid to the next along the chain—the peptide bond—was stiff, so that the atoms at each end of it had to lie in the same plane. The alpha helix was generated by this restriction.

But the interconnectedness of science is itself the most stringent restriction on the making of new theory. A new theory, in replacing a successful older one, at the very minimum must account for all the results that the old one explained, and at least as well as the old one did. Einstein did not *overthrow* Newton—he replaced Newtonian theory, the Newtonian program in its most advanced form, after a century and a half of unrivaled success, with a new theory that included the Newtonian and went on from there. This rigorous restriction on theorizing was first formulated explicitly by Niels Bohr, in the early days of quantum mechanics. Bohr called it *the correspondence principle.* It served him as an instrument—a "magic stick," he said—in the search for new laws. The correspondence principle has been compared to the Hippocratic maxim in medicine: *First of all—do no harm.* The correspondence principle applies throughout science and mathematics. It stops arbitrary speculation. It prevents the loss of what has already been achieved. Therefore, it guarantees as can be guaranteed nowhere else in life that in science, in its own domain, progress is real. The revolutions of science have deeply, radically, subtly altered our sense of ourselves and of the world around us—our sense of what Lucretius called, simply, the nature of things. It's a truism to add that the revolutions of science have enabled enormous changes not only in how we think of the world but in the way we live (and equally a truism to observe that many of *these* changes are hardly progress). For all these radical consequences, though, the revolutions of science are profoundly conservative. The necessity embodied in the correspondence principle makes them so.*

I asked Dirac when physics will end—what problem, if solved, would achieve the final unification so that all that would remain would be the working out of the details.

"I don't think one can answer that," he said. "We don't know: we are going ahead into an unknown region, and we don't know what it will lead to. That makes physics so exciting."

He paused. I waited, then asked if he still felt that excitement.

"Oh, yes. Yes!"

What was the relation between theory and observation?

"Oh, I think that's just our original question: it's most important to have a *beautiful* theory. And if the observations don't support it, don't be too distressed, but wait a bit and see if some error in the observations doesn't show up."

---

*The best discussion of the correspondence principle in English is a small, rather technical, hard-to-find, and scandalously overpriced book by a Polish logician: Wladyslaw Krajewski, *Correspondence Principle and Growth of Science* (Dordrecht and Boston: D. Reidel Publishing Company, 1977). A neat short summary is the review of Krajewski's book, by Philip Morrison, in *Scientific American*, November 1977.

How does one recognize beauty in a theory?

"Well—you feel it," Dirac said. "Just like beauty in a picture or beauty in music. You can't describe it, it's something— And if you don't feel it, you just have to accept that you're not susceptible to it. No one can explain it to you. If someone doesn't appreciate the beauty of music, what can you do? Give 'em up! I've found, during the recent celebrations of Einstein's centenary, that Einstein had very much this same point of view."

Where do theoretical ideas come from?

Dirac paused again, then said quietly, "You just have to try and imagine what—the universe—is like."

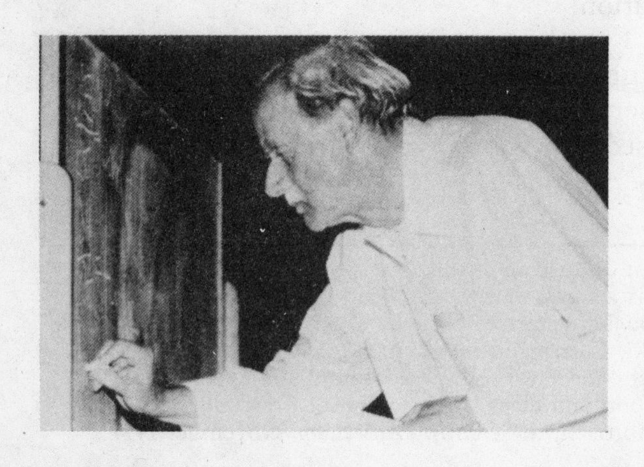

Epilogue: Eight Problems in Search of Solutions

Some problems in science seem impregnable yet yield, some seem fully ripe and yet resist for years. We've pointed out the paradox: the content of tomorrow's science cannot be predicted today. But the paradox reaches further: often in the past the most interesting problems of one period could hardly have been perceived, could not have been correctly, usefully formulated, in the patterns or terms of the period just preceding. Nonetheless, the scientist who wants to do important work will pick an important problem. Among the important problems that can be stated today are these:

The cosmological question: What is the origin of the universe? The "big bang" is the ruling idea today, yet many physicists are disturbed, even angered, by the notion that the universe had a definite beginning before which there was—what? And the evidence for the big bang, though strong, has problems and counterevidence. To replace the big bang might require something as revolutionary as finding an improved version of Einstein's general theory of relativity, which is also a disturbing—and extremely difficult—notion. Even if the universe did begin with a singular event, subsidiary questions remain. For example, Is the universe open, meaning that it's expanding forever? Or closed, meaning that eventually it will all start to move back together again, perhaps to collapse to nothing in another singularity—the little antibang, or soft plop? The answer depends in part on whether there is enough matter—enough mass—in the universe for gravitational attraction to overcome the expansionist momentum eventually. One theoretical physicist said to me, "You can't even talk about these things sensibly until you have a theory of gravitation—a quantum theory of gravitation." That's an example, at least, of the way unsolved problems may get redefined.

A unified theory of the four kinds of forces. Theoretical physicists speak of four forces in nature which are varyingly important at different levels of scale. At one extreme is the strong force, operating at the inconceivably small distances and short times of the interactions of particles within the nucleus of the atom; then comes the weak force, which operates at much longer distances and times—say, at a thousand-billionth of a second or so—between certain particles interacting outside the nucleus; next is the electromagnetic force, which among other things attracts electrons around the nucleus to form atoms, and which is responsible for most of the familiar phenomena of physics and engineering; and finally we know the gravitational force, which in fact is weakest of all and is most important at very large distances—eventually, cosmological distances. The physicists of the 1970s achieved a statement of the properties of the weak and the electromagnetic forces that settled that both were expressions of the same principles. They were reaching with considerable confidence for a unification of the strong force, too, with the weak and the electromagnetic. By the end of the decade some thought they saw, dimly, that all four, including the gravitational, could be brought together under one set of equations. If that is achieved, several physicists have said, physics will be over: the unification of the four forces is as fundamental a statement of the laws of the world as they can conceive. The unification will require, among other things, a theoretical step that could well be described as "a quantum theory of gravitation." With that, one can see that a satisfactory statement of the fundamental laws of physics in such unified terms might well bring with it such an enlightenment that the cosmological question would also fall, as it were, in passing.

The origin of the solar system. A pendulum keeps swinging in this problem. In the nineteenth century, Pierre Simon de Laplace suggested that the present sun and planets originated because a primeval cloud of gas, rotating, slowly contracted and heated up by gravitational attraction, leaving parts of itself behind, swirling as eddies, to condense as the planets. But it was hard to see why the eddies should swirl and concentrate into planets. By the 1930s, the ruling idea was that the planets originated when a star similar to our sun passed so close that great hunks of the sun itself were pulled out in a long streamer—and this matter cooled and collapsed into the planets. By the 1950s, the improbability of such a near passage, and on the other side a refinement of the mathematics by which eddies and vortices could be analyzed, made an improved version of the rotating, contracting, heating cloud of gas the favorite scheme once more. Yet doubts still nagged about those vortices. In the mid-1970s, the notion of an influence from another star revived. This time, however, the rotating, contracting, heating cloud of gas remained—but the emergence of the planets from it was supposed to have been precipitated by a sudden, violent shower of energy and particles from the explosion nearby of a supernova. Evidence that can be interpreted as showing the presence of matter from a different star has recently been found in some meteorites. Claims are bold, the problem still open. Upon its solution hangs much else. For example, if the cloud of gas did not need help from outside to form planets, then a high proportion of other stars in the universe very likely have planets, too, and a high proportion of those solar systems, in turn, may have one planet so constituted and placed that life could develop upon it—suggesting that we are not alone. On the other hand, supernova explosions are rare events—suggesting that intelligent life is rare in the universe.

The origin of life. In 1953, Stanley Miller, a student in chemistry of Harold Urey's at the University of Chicago, took a flask containing a mixture of simple substances, dissolved in water, that might have been present in the atmosphere and seas of the primitive earth before life began. He passed an electrical discharge continually through the flask for a week, to simulate, in a rough-and-ready way, the lightning that would have been frequent on that early earth. When he opened the flask and analyzed his model ocean, he found sizable concentrations of several amino acids—the chemical subunits from which proteins are made. This simple experiment began the subject called origin-of-life chemistry. A great deal has been attempted and written about the subject since, yet a quarter-century later, Miller's experiments remain the most important demonstrations that he or anyone else has yet achieved. The perception of the problem has progressed faster than its solution. Many types of chemicals are essential to life, but the most characteristic and the most important are two kinds—proteins and nucleic acids. Proteins are very long chains of amino acids; nucleic acids are very long chains of components called nucleotides. The exact and specific sequence of different kinds of nucleotides in a string of nucleic acid constitutes a message in code, the genetic information, that dictates the sequence of amino acids in a corresponding protein. A protein must have a nucleic-acid molecule to specify its sequence; a nucleic-acid molecule in order to contain the correct sequence must be formed by a matching-up process on another nucleic-acid molecule already in existence. The subunits of the two kinds of chains can, indeed, assemble themselves in chains simply at random—up to a length of, say, thirty-odd pieces. Above that, the random processes break down the chains as fast as they build them up. In life, long chains of either sort are assembled with the help of structures and chemicals already existing in the cell—including catalysts, certain enzymes. Enzymes are protein molecules. The way through the paradox has perhaps been sighted by Manfred Eigen, in Germany, who, though a physical chemist, turned to biophysics several years ago. He proposes a complex system of interacting cycles that might allow partially random, preliving processes to put together, say, a nucleic-acid chain of useful length. But Eigen's ideas are abstract, and the experiments to test them are not easy to imagine. The question of the origin of life remains one of the most fascinating in science.

The quantitative basis of natural selection and evolution. The fact of the evolution of species is not in doubt. The theory of their evolution by the great two-stroke process, random mutation creating varieties of organisms that are then acted upon by natural selection, has one serious defect—it is, as yet, able to say little about the rates of evolutionary change. Measurement of these rates is still difficult. It requires, to begin with, methods for comparing genetic messages—which means finding the exact sequence of the nucleotides in each gene, to see, point by point, how far one organism's genetic message has diverged from the ancestors it shares with its nearer and farther relations. Methods of sequencing nucleic acids have been developed only in the last five years. Already, so many sequences have been determined that the only practical way to store and compare them is with large-capacity computers. In this way, we will begin to know, more precisely than any classification of features can tell us, how far organisms have traveled from each other. That still won't tell us how fast they are changing. Some experiment is possible, with organisms—bacteria, molds, yeasts, certain insects—that live many generations in a few months or years, so that we can try to find how fast they can be driven to evolve. Theoretical studies will also begin to be possible.

Development and differentiation—or how the fertilized egg becomes the organism. In the quarter of a century ending in about 1970, molecular biology was triumphantly successful in unraveling the intricate interplay of events in the living cell, the processes of heredity and growth that comprise the secret of life. That is, they were successful in doing so for the simplest single-celled organisms—bacteria and the viruses that prey upon them, and certain algae. At that time, many molecular biologists thought it should be easy to verify that these processes—the secrets of life—are fundamentally the same in higher organisms. It became evident, though, that although the processes in higher organisms are similar they are immensely more complicated. The reason is basic: creatures above the bacteria grow from one single cell—typically, the fertilized egg—to a multicelled organism in which various cells have different, specialized functions. This is the problem of embryology. The central question in biology today was posed by William Harvey nearly 350 years ago: *omnes ab ovo,* everything comes from the egg. Development and differentiation of the organism from the potential contained in the single cell requires, among other things, control mechanisms to turn genes in one cell on and off in response to, among other things, signals from other cells of similar or different kind. These controls are stupefyingly complex; the molecular biology of higher organisms is turning up delightful surprises. Biologists were baffled and worried in the mid-seventies; more recently, though, they have regained their buoyant assurance that the problems are tractable.

Aging and death. For higher organisms, the price of youth is the inevitability of old age. Although the organism is able to grow to maturity by means of the control mechanisms of development and differentiation, at some point these mechanisms begin to fail and the organism no longer is able to keep itself up to the standard of the specifications in its genes. It has been suggested that the failure may be strictly a matter of chance—that is, of the inevitable accumulation of errors, mutations, in copying the genetic message as cells in the body are replaced by daughter cells and these in turn replaced. This accumulation of what cyberneticists call noise may reach the point where the cellular mechanisms break down in what Leslie Orgel, at the Salk Institute, has described as an error catastrophe. If this theory of aging is correct, the fundamental process may be intrinsically without remedy. Medical research, on a broad front including preventive medicine, would then have the residual task of combating, one by one, the various disorders that cause *premature* aging—of itself, of course, a task of highest importance. Yet some biologists hope that, as the mechanisms of control are sorted out in the molecular biology of higher organisms, as yet unimaginable possibilities of preventing or repairing the failures of aging will emerge. What seems sure is that the prevention and cure of particular diseases that come with age, like arthritis or the cancers, will appear as by-products of the research into the fundamental processes at the beginning and the end of the curve of development and degeneration that describes our lives.

Neurobiology. Many among the biologists who had, as they thought, pretty well solved the secrets of life at least in single-celled creatures turned, in the late 1960s, to what seemed then the last biological frontier: the nervous system. They hoped it would yield to the combination of genetics and physical, structural chemistry that characterized their successes in molecular biology. Small, encouraging discoveries were made. Breakthrough proved elusive. By the end of the seventies, the problem was stuck fast. How animals perceive, how they recognize pattern, how their nervous systems are organized to process information—and whether certain kinds of information are in fact built into the nervous system—are if not the deepest certainly the most complex problems that can be posed in science today. Nonetheless, on the way—even now—we are elucidating processes of crucial importance to controlling such things as pain and the addictions and to understanding madness. Yet recall John von

Neumann's warning that a complete description of how we perceive may be far more complicated than this complicated process itself—that the only way to explain pattern recognition may be to build a device capable of recognizing pattern, and then, mutely, point to it. *How we think* is still harder, and almost certainly we are not yet breaking this problem down in solvable form.

Science has come a long way in five hundred years. Yet it is only starting. Our list of eight problems in search of solutions begins with the origin of the universe, goes on to the foundations of physics, the origin of the solar system, the origin of life, the basis of evolution, the process of individual growth and aging and death, the means of perception and the nature of thought. In short, what's still to be learned is the history of the universe, the world, and ourselves.

Bibliography

I. Some Classics

Lucretius, *The Way Things Are—The* De Rerum Naturae *of Titus Lucretius Carus,* translated by Rolfe Humphries (Bloomington and London: Indiana University Press, 1968). Lucretius was the first poet of realism and so of science. His epic is austere, panoramic, cumulatively magnificent—and early in the history of Western thought it set up the radical idea, which he got from the philosopher Democritus, that all the variety we perceive in the world is the result of the stupendous multiplication and endless permutation of one or a few kinds of invisibly tiny particles, called atoms.

William Kingdon Clifford, *The Common Sense of the Exact Sciences,* edited by Karl Pearson and James R. Newman, with a preface by Bertrand Russell (New York: Dover Publications, 1955). Common sense is the basis of everyday prudent judgment, and in the sciences such judgment is rooted in the quantitative manner of thinking. Clifford was a brilliant English geometer of the latter half of the nineteenth century. He was also a gifted popularizer. His book is a classic introduction to the principles of mathematics—number, space, quantity, position, motion—for the nonmathematical. Bertrand Russell read it with enthusiasm as an adolescent, and in his old age reread it with respect and with the highest praise for its clarity.

D'Arcy Wentworth Thompson, *On Growth and Form,* abridged edition edited by John Tyler Bonner (Cambridge and New York: Cambridge University Press, 1961). The book we all go back to, as in chapters 2 and 3 of *The Search for Solutions.* Bonner's edition trims away the truly out-of-date, then with excellent notes brings the rest up to date.

G. H. Hardy, *A Mathematician's Apology,* with a foreword by C. P. Snow (Cambridge and New York: Cambridge University Press, 1967). In ninety pages of small format, Hardy creates in the reader the particular sensations of thinking mathematically, so that one leaves the book startlingly clear-headed, knowing directly what this kind of beauty consists in. Hardy has transcended popularization: this is the thing itself, even if not in all its intricacy. Snow's memoir of his friend, nearly as long as Hardy's text, is another sort of surprise, deft and loving.

Norbert Wiener, *Cybernetics: Or Control and Communication in the Animal and the Machine,* second edition (Cambridge, Mass.: MIT Press, 1961). I describe the historic influence of this book in chapter 5.

Peter B. Medawar, *Pluto's Republic* (Oxford and New York: Oxford University Press, 1982). The complete collection of Medawar's essays on science and scientists, written from the inside and with wit, elegance, and brilliance.

II. History

Martin Gardner, *The Ambidextrous Universe: Left, Right, and the Fall of Parity,* second edition (New York: Charles Scribner's Sons, 1978). The most fundamental pattern in nature is its symmetry with itself; if that symmetry were unbroken, Alice could never tell when she was in the looking-glass world. Hence the most exciting discovery in physics so far in the second half of our century is that, at the subatomic level, symmetry is not perfect: the universe exhibits a difference between right and left. Gardner's history of this discovery and its implications was written for the general reader and is accomplished and unpretentious.

Abraham Pais, *'Subtle Is the Lord . . .': The Science and the Life of Albert Einstein* (Oxford and New York: Oxford University Press, 1982). Where did Einstein's ideas come from? To show that, Pais is rigorous and in places necessarily mathematical. As close to a definitive intellectual biography of Einstein as is likely to be possible: as the subtitle suggests, the life is subordinated to the science, but it is not squeezed out. The main title is the first phrase of a celebrated remark of Einstein's, originally in German, of which the other half is, "but he is not malicious."

Horace Freeland Judson, *The Eighth Day of Creation: Makers of the Revolution in Biology* (New York: Simon & Schuster, 1979; London: Jonathan Cape, 1979). A history of the chief discoveries of molecular biology, told by means of a series of discussions with and interlocking biographical sketches of the scientists themselves—Francis Crick, James Watson, Linus Pauling, Jacques Monod, François Jacob, Max Delbrück, Max Perutz, and a hundred others. My aim was to reconstruct the circumstances, psychological and intellectual, of each discovery—to capture the moments and movements of understanding.

Ernst Mayr, *The Growth of Biological Thought: Diversity, Evolution, and Inheritance* (Cambridge, Mass., and London: Harvard University Press, 1982). Physics has long been the very model of what a modern science ought to be; but biology is different from physics, far more complex, uniquely hierarchical, and inescapably contingent upon the history of life with its accidents as well as its necessities. Mayr, one of the grand elders of evolutionary theory, argues passionately in his opening chapters for the epistemological autonomy of biology, then demonstrates with encyclopedic erudition what that autonomy has meant in the development of the science.

III. Philosophy

Karl Popper, *Unended Quest: An Intellectual Autobiography* (La Salle, Ill.: Open Court Publishing, 1976; London: Fontana/Collins, 1976). Bryan Magee, *Popper* (New York: Viking Press, 1973; London: Fontana/Collins, 1973). Popper is the most important philosopher of science of the modern day—for what he has cleared away, for what he has made clear, and for the many others he has stimulated to drive his ideas further. Popper's autobiography, together with Bryan Magee's trim and expert introduction, provides the best way in.

Imre Lakatos, *The Methodology of Scientific Research Programmes*, edited by John Worrall and Gregory Currie (Cambridge and New York: Cambridge University Press, 1978). Lakatos was a philosopher and logician, born, educated, and then imprisoned in Hungary, whence he escaped to England in 1956. He died in 1974, aged fifty-two. By that time, he was vigorously developing the idea that vital, productive science is characterized not exactly by the individual great theories that get general public recognition but rather by the bundles of theories, major and ancillary, that attract and focus the work of many scientists and that, accordingly, he called scientific research programs. I rely on this idea in chapters 8 and 9. At Lakatos's death, his speeches and papers had been published in scattered and specialized places; two colleagues have gathered the most important into a pair of small volumes, of which this is about method and the other mostly about mathematical logic.

Ian Hacking, *Representing and Intervening* (Cambridge and New York: Cambridge University Press, 1983). Scientific observation is not the passive recording or interpretation of what's out there ("representing"), Hacking says, but active interference, the isolation, indeed the creation of the phenomena ("intervening"). We get at the natural world by disturbing it. Hacking is the most original of the younger generation of philosophers of science, and the best writer among them, in a muscular, sometimes knotty way. He appears to be working towards a robust new realism, and this book might best be read as a first draft in that enterprise.

John Ziman, *Reliable Knowledge: An Exploration of the Grounds for Belief in Science* (Cambridge and New York: Cambridge University Press, 1978); *The Force of Knowledge: The Scientific Dimension of Society* (Cambridge and New York: Cambridge University Press, 1976). Ziman, himself a physicist, writes about the ways scientists work, think, interact. He does so with intelligence, energy, and strong, helpful analogies—as well as great good sense in opening up complexities.

Index